ADVANCES IN MANUFACTURING ENGINEERING AND TECHNOLOGY

ADVANCES IN MANUFACTURING ENGINEERING AND TECHNOLOGY

M. Adithan
Director and Senior Professor of
Mechanical Engineering
Vellore Institute of Technology University
Vellore, Tamil Nadu

PUBLISHING FOR ONE WORLD

NEW AGE INTERNATIONAL (P) LIMITED, PUBLISHERS

New Delhi • Bangalore • Chennai • Cochin • Guwahati • Hyderabad
Jalandhar • Kolkata • Lucknow • Mumbai • Ranchi
Visit us at www.newagepublishers.com

Published by New Age International (P) Ltd., Publishers
First Edition: 2010

Branches:

- No. 37/10, 8th Cross (Near Hanuman Temple), Azad Nagar, Chamrajpet, **Bangalore**-560 018. Tel.: (080) 26756823, Telefax: 26756820, E-mail: bangalore@newagepublishers.com
- 26, Damodaran Street, T. Nagar, **Chennai**-600 017. Tel.: (044) 24353401, Telefax: 24351463 E-mail: chennai@newagepublishers.com
- CC-39/1016, Carrier Station Road, Ernakulam South, **Cochin**-682 016. Tel.: (0484) 2377004, Telefax: 4051303 E-mail: cochin@newagepublishers.com
- Hemsen Complex, Mohd. Shah Road, Paltan Bazar, Near Starline Hotel, **Guwahati**-781 008. Tel.: (0361) 2513881 Telefax: 2543669, E-mail: guwahati@newagepublishers.com
- No. 105, 1st Floor, Madhiray Kaveri Tower, 3-2-19, Azam Jahi Road, Nimboliadda, **Hyderabad**-500 027. Tel.: (040) 24652456, Telefax: 24652457, E-mail: hyderabad@newagepublishers.com
- RDB Chambers (Formerly Lotus Cinema)106A, 1st Floor, S.N. Banerjee Road, **Kolkata**-700 014. Tel.: (033) 22273773, Telefax: 22275247, E-mail: kolkata@newagepublishers.com
- 16-A, Jopling Road, **Lucknow**-226 001. Tel.: (0522) 2209578, 4045297, Telefax: 2204098 E-mail: lucknow@newagepublishers.com
- 142C, Victor House, Ground Floor, N.M. Joshi Marg, Lower Parel, **Mumbai**-400 013. Tel.: (022) 24927869 Telefax: 24915415, E-mail: mumbai@newagepublishers.com
- 22, Golden House, Daryaganj, **New Delhi**-110 002. Tel.: (011) 23262370, 23262368, Telefax: 43551305 E-mail: sales@newagepublishers.com

---

ISBN : 978-81-224-2674-8

Rs. 495.00

C-10-02-4331

Printed in India at Glorious Printers, Delhi.
Typeset at Innovative Processors, Delhi.

PUBLISHING FOR ONE WORLD
NEW AGE INTERNATIONAL (P) LIMITED, PUBLISHERS
4835/24, Ansari Road, Daryaganj, New Delhi-110002
Visit us at **www.newagepublishers.com**

Preface

Manufacturing is an exciting and rewarding discipline that has significant impact on a society's standard of living and economic independence and prosperity. Manufacturing programmes have created leaders in the industry.

Students in manufacturing learn creative and analytical skill that will enable them to quickly diagnose and solve manufacturing problems with insight from both engineering and management perspectives. They also develop interpersonal and communication skills that will prepare them to work as a part of an engineering team and effectively interact with vendors, management and production personnel. In addition, they receive hands-on training in industries and modern laboratory facilities and learn to use computers to design, analyse, implement and control manufacturing operations.

There is an increasing demand for manufacturing professionals who are knowledgeable and skilled in the management methods, technologies and equipment and tooling needed to produce quality and affordable products and services. Such individuals must also be able to effectively coordinate the procurement, installation and start up of production operations as well as improve the productivity of existing operations. Few professions encompass such a broad range of activities and utilize so many skills.

Career opportunities in manufacturing are plenty and rewarding. Leaders of industry often have manufacturing background. Typical entry-level job titles include manufacturing engineer, quality manager, process engineer, tool engineer, product engineer, production engineer and technical sales engineer/manager. New graduates are typically employed in the technical positions but have the opportunity to move into management positions as well.

The job outlook for manufacturing graduates is bright and should continue to be strong into the future. When one considers that everything that doesn't exist as part of nature is the product of some form of "manufacturing", it is easy to see that manufacturing is an integral part of our society and generates an "ever-growing work force". Progressive companies and industries worldwide are always on the look out for qualified individuals who can provide leadership in improving the quality and productivity of their manufacturing operations.

A comprehensive knowledge and study of the advances made in the areas of manufacturing management, methods, technologies, equipment and tooling will be useful to the manufacturing engineers and technologists in building up their career prospects. This book is a step in this direction.

M. Adithan

Contents

1

Global Manufacturing of Ultra-Class Equipment

Stephen R. Light*

Here we describe how Bucyrus International of Milwaukee Wisconsin in the United States is preparing itself to face the new millennium. Established in 1880 in a small town in the state of Ohio, U.S.A., Bucyrus, and known for most of this century as Bucyrus Erie, the Bucyrus International Company is a multinational manufacturing and service company focused on supporting the global surface mining industry. We are proud to be alliance partners of Bharat Earth Movers Limited, Bangalore, in the project to construct three drag-lines for the Northern Coal Fields of Coal India. Bucyrus manufactured surface mining equipment includes drag-lines, electric rope shovels, blast hole drills, corresponding bucket and dippers to move the earth and long-life spare parts. Our product line transcends machinery and in some countries we actually operate draglines for the owners of the mines. Today, more than 1,800 Bucyrus employees serve in more than fifty countries supporting 1,127 operating machines. Bucyrus has designed and produced more than 90 per cent of the world's active draglines. Our design and development expertise is unmatched for experience, quality and efficiency, as are our products.

The surface mining industry is the source of most of the energy rich coal produced in the world today, as well as majority of the copper, gold, diamonds, iron ore, phosphates, silver, zinc and many other minerals needed by all developed nations. Of the 7,015 presently identified surface mines in the world today, we believe that approximately 445 are of a size that may economically employ equipment of the scale manufactured by Bucyrus International. The remainder are typically very small and serve only local markets. Many of our ultra-sized machines are in use in India today.

While mining in the United States is a well developed industry with a long history, it is also a mature industry with little new exploration and little resulting growth for mining equipment manufacturers. However, surface mining throughout the remainder of the world is rapidly expanding. In general, it is growing in direct relation to the improving standards of living and increasing populations of the nations of the globe. This world-wide growth in mining related equipment manufacturing is driven by the need for energy and increased agricultural production, as exemplified by India's increasing demand for coal for electrical power generation, and for phosphates for fertiliser necessitated by continued population expansion with stable cultivated acreage.

As a result of these marketplace realities, Bucyrus finds itself in a situation where seventy per cent of its products are sold to customers outside its home country. This shift to an export dominated customer base has had a direct bearing on Bucyrus' manufacturing strategies.

*Chief Executive Officer, Bucyrus Europe, Lincoln, England.

The significance of transportation and geography on Bucyrus' business can best be demonstrated by providing an overview of the company's products. A modern electric mining shovel, often called a "Rope Shovel", weighs approximately 3,000,000 pounds, or 1,650 metric tons, when fully assembled, stands 100 feet or greater in height, and 150 feet in length. A large sized drag-line, such as those being built presently to expand Coal India's Northern Coal Fields fleet, may weigh as much as 5,900 metric tons stand 70 m or more high, and have a length of 140 m. In both cases, components that comprise the machines will generally weigh in excess of 50 T and will be very difficult to transport. In fact, most countries in which the machines are ultimately employed have road weight restrictions that constrain Bucyrus' ability to deliver components in the sizes that are economical to build. For example, in the supply of machines to northern Canada, the company must be continuously aware of the depth of the ground freeze during the winter and adjust trucked loads accordingly.

The size, weight, and complexity of these ultra large sized machines, together make transportation a major cost element in the manufacture and assembly of the unit. These limitations frequently have a substantial negative impact on the price the customer is charged for the equipment.

An additional consideration in the development of the manufacturing plan for such machinery is the expectation that the machines will be in service for more than twenty-five years. Life cycle management that is the care of the machine throughout its useful economic life must be carefully planned jointly between the manufacturer and the customer. Bucyrus' capability to develop indigenous maintenance support is therefore a major positive consideration during the purchasing process.

In recognition of the ultimate customer's requirements, as well as the physical constraints presented by the size of the machinery that Bucyrus International produces, the company has developed and implemented a global manufacturing strategy. This strategy has proven to be cost effective and beneficial to the end user reducing the cycle time from order to machine receipt. An additional benefit of Bucyrus' global manufacturing strategy is the production technology transfer resulting from the introduction of specialized manufacturing processes to the workforce of the producing location or country.

In this paper, I would like to share Bucyrus' global manufacturing strategy and discuss how the company has configured its internal processes to optimise customer support through appropriately distributed production planning.

The heart of the production process at Bucyrus is our new "Enterprise Wide Resource Planning" (ERP-II) System. The new information technology system replaces the one developed by Bucyrus International in the earlier stages. The old system employing the first generation process, a classical Materials Requirements Planning system (MRP-I), was mainframe based, and focused on management of the material acquisition and factory shop floor activities within the company. This old system served the company for more than twenty years with mixed success. While the software itself was robust, it was also very restrictive in terms of improving customer service levels, promoted unnecessarily high inventory levels, and was not capable of managing the mixed demands of a complex original equipment and after-market parts business operating on six continents. It was totally lacking in any ability to perform the iterative "What If" analysis that are necessary today to optimise a company's performance. Our new system is user friendly, client server operated and windows based.

While the most efficient and expensive new information technology systems offer extraordinary capabilities, the most significant challenge facing the implementing company is not learning how to perform specific transactions on the computer screens, but rather how to incorporate the available tools the system contains in the management processes of the company. To do this, Bucyrus has spent considerable amount this year in training its workforce in modern materials management, project management and product development techniques. In order to facilitate the normally difficult transition from one system to another, our new system has actually been "live" at our employee's desktops for many months prior to the conversion date. By providing this access in an off-line manner, employees can explore the new environment on their own, much the way most of us learn new computer software. The system conversion has become less threatening and more efficient.

Now that we are fortified with this new information technology tool, we are enhancing our structured management process to enable management to accelerate our factory throughput, reduce cycle times and identify and prevent defects. Our unique approach to "team managing" the business is one of the key elements in our globalised manufacturing strategy.

Four major factors have been recognised in order to respond to the globalisation of our market place. They are:

1. The recognition that Bucyrus is in the project business and that our machines will of necessity be built to a customer specification rather than from a pre-developed pattern.
2. Many countries have adequately developed technical and physical infrastructures to produce major components for our equipment.
3. Indigenous manufacturers may provide a substantial cost reduction potential thereby lowering end user prices and improving competitiveness.
4. The optimum means by which to assure long term satisfactory life cycle performance of the machines is to have support facilities and personnel as closely proximate to the mines and machines as possible.

Our global manufacturing decision process begins with a careful review of the customer's request for quotation immediately upon receipt. Of course, the physical performance specifications are understood, but at this point we also analyse and begin development of the production plan as well. Following a thorough review of the customer's documentation a comparison is made to determine the optimum balance between labour costs, transportation and in-country support requirements. If the economics and logistics support a decision to manufacture away from the United States, then we set about finding an appropriate partner with whom to work. We seek companies that are able to satisfy most if not all of the above-identified criteria with one key additional requirement. The company with whom we will form an alliance must have a management process that is complimentary to ours, with open lines of communication, and a focus on long term rather than short term performance and customer relationships. Our partner is BEML, Bangalore, in India.

An excellent example of the cooperation and benefits such a global alliance can provide are exemplified by the contract BEML has received for the provision of three W2000 Bucyrus draglines to Coal India. The largest structures of these machines are being built in India by BEML utilising drawings and technical support provided under license by Bucyrus' United Kingdom office. Subcontractors to Bucyrus, UK are organizing manufacturing of the gears, shafts, pulleys and other critical running gear to the required metric specifications, while in the United States, Bucyrus North America is providing the electrical drives, motors and controls, as well as overall project, and management for the Bucyrus portion. Direct contact with the end user is maintained by BEML with frequent participation by whichever portion of Bucyrus may best respond to customer needs. Assuring the coordination of the Bucyrus portion of the work is well integrated to that of BEML and is enhanced by the Bucyrus employees who reside permanently in Bangalore. This presence facilitates rapid and understandable communication with the UK and Milwaukee, Wisconsin, U.S.A. operations and serves to resolve time zone differences that would otherwise delay decision making.

Of particular concern on a project, such as this one, is the necessity to integrate production schedules between the numerous manufacturing sites and the ultimate erection location. This is accomplished through a series of scheduled meetings and the use of purchase orders with defined deliverables and schedules. There is a natural tendency during such a lengthy process for the end user and the suppliers to make changes to the original specifications in order to enhance the long-term performance of the end products. Such changes may occur at anytime during the development and erection of the machinery. Close coordination of the numerous suppliers with the end user can facilitate this effort without any impact on schedules. Once again, the importance of having the proper partner and having predetermined communication vehicles is key to customer satisfaction.

Numerous other advantages accrue to the host country when their equipment supplier employs a **local manufacturing strategy**:

1. Lead times for machinery are reduced; hence working capital investment is reduced.
2. Local maintenance capabilities are developed.
3. Low cost local suppliers are developed to provide on-going support for the equipment throughout its anticipated life cycle.
4. Technology transferred to local manufacturing companies may be of value to other areas of their business and the general economy of the country. An example of this advantageous technology transfer is the introduction of the option of new Siemens Electrical Drives and motors to the W2000 draglines for Coal India. These drives and motors are applicable to many other processes outside of the draglines on which they are being installed. Steel mills, rock crushers and other machines will be improved when the technology migrates to them.
5. Local skills are enhanced in the areas of procurement, project management and quality improvement.

Finally, a key element of Bucyrus International's global manufacturing strategy is the recognition that our people cannot adequately assist our alliance partners from Milwaukee, U.S.A. but rather they must be willing and able to rapidly "go where the action is". Our reliance on the quality and commitment of our people and their wealth of historical knowledge and experience are at the very foundation of our strategy. It is, after all, a unique competitive advantage that our company offers to its customers and its manufacturing partners throughout the world.

The decision to "go global" has been supported by Bucyrus' equipment, systems and people. This unique combination of resources and our carefully selected partners throughout the world, global manufacturing would no longer be a dream but a reality.

2

The Future of Competitive Manufacturing

*Dr. Paul M. Swamidass**

The future of manufacturing technology is tied closely to information technology. *Fortune* magazine publishes a list of the largest 500 businesses in the U.S., it is called the *Fortune 500*. I recently read that 70 per cent of the original 500 firms are no longer listed there. What happened to them? Many became unsuccessful and went out of business; downsized; or merged. Further, many new firms that emerged later have become larger and overtaken those making up the original 500. Technological developments are substantially responsible for the re-making of the *Fortune 500*.

New technologies threaten established businesses, or create new businesses. Example, the minimill technology (mini-steel plants) for steel production has captured 40 per cent of steel production in the U.S. from basic steel producers.

In the mid-seventies, the CEO of a large computer company in the United States is quoted as saying, "Why would anyone want to use a computer at home?" He could not visualize the personal computer's (PC) role. However, as a result of the introduction of personal computers, Microsoft became one of largest firms in the world in terms of market capitalization in less than 20 years. Thus, technological progress provides threats as well as opportunities.

Lesson: It is hazardous to make predictions about technology in this era. It is estimated that during the last 30 years, 90 per cent of all scientific knowledge was generated, and 90 per cent of all scientists, whoever lived, are now living and working. Consequently, we can expect more technological revolutions in the future.

Our Indian industry has been outstanding in its competitiveness in the last ten years, In the ten years between 1989 and 1999, software exports from India grew by 2500 per cent. Software exports were US $2.6 billion in 1999. That is, the exports in 1999 were 25 times larger than the exports in 1989. Further, the exports in 1999 were more than eight times the exports five years earlier in 1994. The growth rate of this industry is truly world-class.

Exports are one of the indicators of an industry's competitiveness. If an industry is successful in increasing exports better than other nations, it means that the industry is working to world-class standards, or it is the world standard for excellence. Indian software industry, which has a strong foothold in Bangalore, has the true credentials of a world-class industry. This industry taps into the abundant supply of educated young men and women in this country. There is generous capital flow from abroad to start new ventures and joint ventures to use this abundant human capital. Many of them are trained in India's engineering colleges.

**Auburn University, Auburn, Alabama, U.S.A.*

If India's manufacturers could only emulate the software industry, India will be a major industrial nation in a short time.

REGAINING MANUFACTURING COMPETITIVENESS: THE CASE OF U.S.A.

The twenty-five years from 1980 to 2005 represent a unique period in the history of manufacturing in the United States. During this period, there was an escalation of competitive pressures on US manufacturers. The competitive pressures manifested in several ways:

1. Higher U.S. wage rates relative to many new Asian exporters.
2. The free access of U.S. domestic markets to exporters from other countries, not vice versa.
3. Rapidly changing product technology in computer and electronics related products required frequent new product introduction, some requiring new product introduction every 6 to 12 months.
4. The rising expectations of customers concerning product quality.
5. The rising expectations of customers concerning customer service.

MANUFACTURING TECHNOLOGIES TO THE RESCUE

In response to the pressures, manufacturers in the U.S. turned to extensive use of inventions and/or imitations of exceptionally sound manufacturing technologies. Manufacturing technologies are classified as hard and soft technologies.

Hard technologies are hardware and software intensive. The examples are CAD, CAM, CNC machines, CIM, FMS, automated inspection, robots, LAN, WAN, and so on.

Soft technologies are manufacturing and production know-how, techniques and procedures. Examples are: bar codes, concurrent engineering, JIT manufacturing, manufacturing cells, MRP, SQC, simulation and modelling, TQM, TPM, and so on. Soft technologies are not necessarily hardware/software dependent.

Many of these newer manufacturing technologies are now an integral part of manufacturing in the United States. As a direct result of the use of these hard and soft technologies, U.S. manufacturers report many improvements.

In a study sponsored by the National Association of Manufacturers (U.S.A.) and the National Science Foundation (U.S.A.), I investigated the use of manufacturing technologies in the U.S. Out of 1025 manufacturing plants participating in the study, more than two thirds of the respondents reported that the benefits from manufacturing technology use included decrease in cycle time, decrease in manufacturing costs, increase in product lines, and increase in return on investment (ROI). They reported that the increase in product line during 1994–1997 was 25 per cent, increase in ROI was 22 per cent, increase in market share was 21 per cent, decrease in manufacturing costs were 11 per cent and decrease in cycle time was 16 per cent.

Over the last twenty years, through the use of various manufacturing technologies, U.S. manufacturers have:

1. Become more profitable
2. Become flexible
3. Reduced production batch sizes without incurring any cost penalty
4. Become cost effective
5. The capability to bring new products to markets quicker

6. Reduced manufacturing lead time
7. Reduced inventories
8. Reduced the number of suppliers
9. Increased supply reliability
10. Increased the quality of the product, distribution system, and customer service
11. Improved the design process to make manufacturable designs
12. Become skilled in the use of teams for problem solving, design and development, and lead-time reduction.

GROWTH IN THE OUTPUT AND EXPORT OF MANUFACTURED GOODS

U.S. manufactured exports have grown steadily since 1987; from $200 billion to $597 billion in 1998. While the gross domestic product (GDP) increased 300 per cent between 1980 and 1987, manufactured exports increased to 368 per cent. U.S. exports of manufactured goods in 1991 were 12 per cent of total world exports, which was 0.1 per cent better than that of Japan. U.S. exports, as a percentage of world exports, were better than Japan by the same margin in 1995. German exports dropped from 14.2 to 12.2 per cent during the same period. Export performance of U.S. manufacturers is one indication of their competitiveness in the global market.

COMPETING IN THE FACE OF GLOBAL MANUFACTURING OVERCAPACITY

The progress in manufacturing in the United States stepped up at a time when developing nations, particularly in the Pacific Basin (Pacific Rim countries), added manufacturing capacity rapidly during the last two decades. China, Taiwan, Hong Kong, S. Korea, Malaysia, Thailand and Singapore, in addition to Japan, expanded their manufacturing capacities substantially since 1970s. The resulting overcapacity created a strong competition for markets among manufacturers. Only the best and most efficient have survived.

In an era of expanding global manufacturing capacity, U.S. manufacturers with relatively high labour wage rates have retained their competitiveness because of the improvements in manufacturing listed above.

STRATEGIC THINKING

In the last twenty years, strategic thinking has overtaken single-minded cost reduction and cost minimization in manufacturing. Consequently, the pursuit of cost, quality, flexibility, dependability and timeliness has replaced the single-minded cost reduction in manufacturing firms, which was the norm in manufacturing until the sixties and seventies. Now, manufacturers find competitive advantage through better design, improved customer satisfaction, quick response, faster new product introduction and other goals overshadowed in the past by the sole pursuit of cost reduction.

Strategic thinking in manufacturing filters down to the lower level in manufacturing when the corporate strategic planning process accepts manufacturing to be an integral part of strategic decisions that affect the competitiveness of the plant. This requires that the strategic business units (SBU) develop and execute an appropriate manufacturing strategy that is consistent with the overall strategy of the business. Once the manufacturing strategy is developed, it enables systematic and timely investment in manufacturing equipment and people to maintain the vitality of manufacturing and keep it competitive.

LEAN PRODUCTION

Lean production, in simple terms, refers to producing more with less. This popular term was introduced by Womack, Jones and Roos (1990) in their best-selling book, *The Machine That Changed the World*. Lean production is the result of the use of several newer manufacturing developments that were adopted by U.S. manufacturers in the last twenty years. The underlying principle is waste elimination in the use of all resources including people and time. While the foundational principles were imported from Japanese manufacturers, they have been Americanized, and smoothly adapted and integrated into U.S. production.

DESIGN REDEFINED

Earlier, product design in the United States was dominated by design engineers' perspective of a product without sufficient input from the manufacturing function, suppliers and customers. Today, concurrent engineering, cross-functional teams, design for manufacturability, and customer inputs in product design are overshadowing the traditional approach to uncompetitive product design.

The increase in manufacturing capacity worldwide has resulted in several new competitors for almost every product. One approach to attract customers in this market is to increase the extent of customisation in products, even in mass-produced products. An important development in this area, called *mass customisation*, increases customisation of mass-produced products. Mass customisation begins with design and is executed by manufacturing.

SUPPLY CHAIN MANAGEMENT REPLACES MUNDANE PURCHASING

The role of suppliers has been greatly increased over the last twenty years. In the earlier part of this century, the tendency of manufacturers was to integrate vertically on the supply side, which ensured captive supply of numerous materials and components needed for assembly. General Motors and Ford Motor Corporation were good illustration of this phenomenon till the seventies. Since the eighties, the trend is to reduce the dependence on in-house supply of most components including technology intensive components. Only the components that represent the core competency of the company are not sourced from outside.

The increased use of outsourcing has escalated the dependence on suppliers. The result is to concentrate on a few good suppliers and integrate them more tightly with the manufacturing operation. The resulting supply chain provides competitive advantage.

HIGHER THRESHOLD FOR QUALITY

The quality of manufactured products across the board has improved over the last 20 years. In many industries, the minimum threshold for quality of the product is very high. Those manufacturers, who cannot meet or exceed this minimum threshold, cannot survive. The automobile and personal computer industries are very good examples of this phenomenon. Statistical quality control (SQC), JIT manufacturing, and total quality management (TQM) principles have been adopted by manufacturers from a variety of industries to produce consistently higher quality products.

Improved and consistent quality is made possible by waste reduction, mistake-proofing (fool-proofing), doing things right the first time, improved control of manufacturing processes instead of inspection of the output, reduction in rework and rejection, improved and frequent operator training, and immediate, collective quality problem solving in quality circles or similar teams.

IMPROVED COSTING AND PERFORMANCE MEASURES

Costing is the process of assigning costs to manufactured products. This process has remained unchanged for several decades, while manufacturing processes underwent dramatic changes. For example, when traditional unit-based costing systems were instituted at the turn of the century, labour input was a major component of the product cost.

A weakness of the approach is the use of direct labour cost to compute overhead costs. Today, with increased automation and labour productivity, in some automated plants, labour costs could be as low as five per cent of the manufactured product cost. When direct labour costs are very low, the overhead cost reaches several hundred per cent of direct labour cost. Consequently, a small change in direct labour cost could alter the total cost very substantially. The negative impacts of this are many. They include the reluctance of the management to invest in equipment and process technology because additional investment would increase the overhead burden rate.

For competitive manufacturing, the investment in equipment and process should be determined by the strategic and competitive priorities and realities. Therefore, traditional costing approaches have been criticized and other techniques, such as activity-based costing (ABC), have received attention in recent years for internal control and management purposes.

Product costs, based on ABC more accurately, reflect the true cost of the product. Decisions based on more accurate costs provide better information for pricing the products, and do not hinder timely investment in automation and equipment.

Additionally, target costing has gained ground as the need for competitive pricing has increased. A product's price is determined using a study of the market, and competitors before it is designed and marketed. Multiple iterations of the design are then undertaken until a viable product could be designed within the targeted price. A well-known example is that of Mercedes-Benz (MB). Until the early nineties, MB cars were built and then priced to cover all the incurred costs and desired profit. After one hundred years of this practice, due to increased competition in the luxury car markets and falling market share, target costing was instituted in this company in the early nineties. Within a few years, MB introduced several new products designed under the target costing principles and recovered.

The M-Class vehicles (sport utility vehicle) designed and produced by MB using target costing has been very successful and profitable.

NEW RESPECT FOR THE CUSTOMER

The customer was an afterthought in traditional manufacturing. Japanese manufacturing principles embedded in JIT manufacturing and TQM have elevated the role of customer satisfaction and customer input into the design and manufacturing of products. Further, customer satisfaction earned through dependable, reliable products and services is now considered an important competitive advantage. The most notable aspect of customer satisfaction is that it cannot be easily and quickly replicated by competition. Manufacturers, who earn reputation for sustained customer satisfaction, will be hard to dislodge from their markets.

COMPETITION IS NO LONGER LOCAL: GO GLOBAL

Increase in global manufacturing and an increased sensitivity to global markets defines the last two decades. The removal and reduction of restrictions to trade and investments have enhanced the global reach of manufacturers across national borders. This was particularly evident in Asia and the Pacific Rim countries.

Global reach of manufacturers makes domestic markets vulnerable to international competition. Manufacturers have responded with globalization of product design, international alliances, and international operations to cope with increased globalization of their markets. The net result is that product prices have remained depressed and manufacturers are forced to be more efficient to remain competitive. Inflation in the prices of manufactured goods has remained well under control in many markets over most of the world. This did not permit manufacturers to raise prices for years to come and kept the pressure on manufacturers to become more productive and efficient.

NEW MANUFACTURING TRENDS

Environmentally conscious manufacturing is changing the way design, raw materials and manufacturing are conducted. Further, the rise of remanufacturing and reuse of durable goods pose new opportunities and threats to manufacturers.

Several new initiatives in manufacturing promise to make manufacturing more flexible and responsive. Agile manufacturing and virtual manufacturing initiatives in the U.S. are aimed at quick response to changing markets and new opportunities. Virtual manufacturing refers to new manufacturing entities created through very rapid integration of scattered resources in one or several firms. The rise of information technologies and the Internet fuels the growth of virtual manufacturing.

Enterprise resource planning (ERP) is a new development. It is essentially a conglomeration of information and decision support systems accomplished by a range of software, which today enables the integration of manufacturers, their suppliers and customers. The promise of ERP is cost efficiency and improved effectiveness in complex organizations, which have multiple facilities spread across many nations. ERP is based on improved data handling, data warehousing, data analysis, and decision making for improved coordination of manufacturing as well as the entire business.

Business process re-engineering, which gained momentum in the nineties is a major force in reducing wasted effort and resources in established manufacturing firms.

For Indian manufacturers, there is much to emulate from the experience of U.S. manufacturers over the last 20 years. While U.S. manufacturers are still the most productive (output per employee) in the world, they have become more effective in terms of customer satisfaction, quality, delivery and new product introduction.

INTEGRATION: THE ROLE OF IT AND INTERNET

Progress and innovations in metal cutting, metal forming, and other manufacturing processes are everywhere. Some of the advances in manufacturing processes will revolutionize manufacturing.

Integration will be a major theme in manufacturing in the years ahead. Integration reduces overhead costs, manufacturing lead times, increases flexibility to change and makes it possible to introduce new products more frequently.

The use of information technology (IT) is a proven way of increasing integration, which results in effectiveness and efficiency of manufacturing firms. Integration technologies are intranet, extranet, LAN, WAN, EDI, and bar codes. These technologies help integrate within and across the supply chain, design functions, distribution facilities, customers, subcontractors, factories, and manufacturing equipment. Manufacturers in developed countries are using more and more integration technologies to become competitive. India is a leader in the information technology field. Therefore, the Internet and its use in manufacturing are relatively new phenomena. The use of Internet for increasing manufacturing efficiency and effectiveness will explode in the near future. The Internet can be used for progressively more and more complex purposes. Internet use by manufacturing firms in ascending order of complexity is listed below:

- Dissemination of company information.
- Dissemination of product information.
- Dissemination of product specifications.
- Internal company or plant communication.
- To process customer request for samples and/or literature.
- Automated sales and order entries through the Internet.
- Bidding and receiving contracts over the Internet.
- Posting of order status on the Internet for coordination purposes.
- Product design and development.
- Training of employees and customers.

The use of the Internet for the above purposes in a manufacturing firm reduces overhead costs. It also contributes to the integration of customers or potential customers, supply chain members, subcontractors, distributors, factories and many other entities. Marketing and administrative costs can be reduced substantially through integration. It makes manufacturers more competitive.

THE FUTURE

Investment in manufacturing technologies requires capital. India's industrial base has a huge potential for growth to meet demands both from inside and outside the country. Domestic capital formation cannot meet most of the demands for capital. Foreign direct investment in manufacturing must be encouraged vigorously. China has outdone India in the area of foreign direct investment. India must seek to become more attractive than any other nation for foreign investors.

According to the Planning Commission, there were 18.7 million unemployed persons in India in 1995. Additional 67 million job seekers will join the workforce between 1998 and 2008. The total non-agricultural jobs in India in 1991 was 67.4 million, out of which 19.6 million were employed in the government: Central, State and Local.

In the past, Indian government created many jobs. Between 1985 and 1991, the government created 1.8 million new jobs but it created only half a million new jobs between 1991 and 1997, and it is expected to decline in the future if the Pay Commission's recommendations for reducing government employment are implemented.

This educated workforce is a resource India must use to expand its manufacturing base in the manner of the software industry, which is now a world-class exporter.

The role of the government is critical for manufacturing to thrive. The government should not be in manufacturing; it should promote manufacturing and create conditions conducive to manufacturing. The government is distracted from its primary purpose by trying to do too many things. The primary purpose is to build a solid infrastructure for businesses to thrive and flourish. The government must do the following to promote manufacturing in India:

1. Develop the infrastructure by investing and attracting investments in power plants, road transportation and communication.
2. Treat literacy as an infrastructure item. Bring literacy rate in the nation to nearly 100 per cent. The national average is about 60 per cent, if you believe the most favourable estimates. In some states, average literacy is around 40 per cent.
3. Attract foreign investors to invest in new manufacturing facilities, which would encourage them to bring their products and manufacturing technologies. There is no faster way of energizing manufacturing in India.
4. Encourage private sectors in India to invest in newest technologies and businesses.

If the infrastructure grows, private investment in factories and manufacturing technologies would multiply, which, in turn, would cause growth in employment and prosperity, which India needs desperately.

3

Manufacturing Industry: Directions for the Future

*N. Ramanuja**

The manufacturing industry is undergoing a paradigm shift. Speed to market, agile manufacturing and virtual corporations—terms firmly rooted in the lexicon of global business—reflect an increasing awareness that competing successfully in the changing contours of global economy would require fundamental changes in the way companies operate.

The new paradigm is the culmination of the continuous transition that began with the industrial revolution in the late 18th century. The concept of mass production, which originated in that era, reached its full development in the 1920s. At that time, it featured three basic characteristics: division of labour, interchangeable parts and mechanisation. Although the pursuit of "perfect interchangeability" remained a challenge throughout the nineteenth century, what was known as the American system of manufacturing became the symbol of industrialisation. As early as 1850, it began to impose itself as the dominant mode of manufacturing and by 1890 it was solidly established in the U.S.A. In contrast, the "European" method relied more on human skills and less on mechanisation or interchangeability, hand-fitting pieces together being a common practice. Craft production, where artisans built products to a customer's specification continued catering to some select market niches.

In the 1970s, the competition intensified among the producers and, at the same time, the variety of products proliferated. As a result, lean production which combines the advantages of craft and mass production while avoiding the high cost of the former and the inflexibility of the latter was born. Faced with a decline in productivity growth, U.S. industries reacted energetically with a variety of approaches as early as the 1970s in Quality of Working Life programmes (QWL), Quality Circles (QC), development of system for planning resources and production (MRP II). In the 1980s and early 1990s there were campaigns for improving productivity, and search for excellence. Projects like Flexible Manufacturing Systems (FMS), robotics, Computer Integrated Manufacturing (CIM), Just-In-Time production (JIT), Business Process Re-engineering (BPR), continuous improvement or Kaizen, Total Quality Management (TQM), Time-Based Competition (TBC) and World-Class Manufacturing (WCM) were mostly attempts by corporations to respond to competition—often foreign. Many achieved only limited success. Their difficulties partly linked to their inability to free themselves from the mass-production paradigm.

At present, while some markets are reaching their saturation and the customers are more demanding, a shift has been taking place from mass production to mass customised production. The concept of mass customisation focuses on satisfying the customers' unique needs with the

**Chairman and Managing Director, Hindustan Machine Tools, Bangalore, India.*

help of technologies, such as agile manufacturing, flexible manufacturing system, computer integrated manufacturing, incorporating explosive changes in computers and communication. This trend has also given rise to virtual enterprise.

The new millennium will witness rapid worldwide political and economic shifts empowering and increasing the number of global competitors. In this border-less world, the diffusion of information and technology will facilitate access to latest know-how and equipment. What were previously Third World countries in Southeast Asia and Latin America are producing sophisticated goods and services. These new competitors are now smarter and more productive. International customers are also more sophisticated and demanding. With access to an unparalleled variety of products from all over the world, they can more easily identify value. As a result, they have become more selective purchasers. They expect quality, reliability and competitive pricing but also want customised products that are delivered quickly. Much of their power lies in shifting product allegiances, which typically focus on goods that provide greater immediate value. These profound changes will render the trends that are witnessed in the manufacturing industry deeper and more pervasive.

Agile manufacturing, the new recipe for success is evolving as a top-down enterprise-wide effort that supports the time-to-market attribute of competitiveness. An agile enterprise could reconfigure operations, processes and business relationships swiftly, thriving in an environment of continuous and unpredictable change.

REASONS FOR VIRTUAL ORGANISATION

In the search of agility companies will rely increasingly on virtual enterprises, virtual manufacturing and virtual reality. There are six strategic reasons for the evolution of virtual organisation:

1. To share the infrastructure cost and associated risk of being in business. One example involves Ford, Chrysler, and General Motors sharing the R and D costs of developing an electric car and battery.
2. To share core competencies. For example, a small job shop specialising in five-axis machining can collaborate with a tool designer to produce a complex mould.
3. To reduce time to market. An example might be Apple inviting Sony to help it with the Power Book and thus reduce its design manufacturing cycle time. This sharing of effort allows a company to avoid putting out the world's greatest product for yesterday's market.
4. To look bigger. Consider one small company that has about 28 employees. Customers don't question the world-class quality of its product when they find that IBM does its manufacturing.
5. To share markets, market access, and market loyalty. A key principle here is cobranding. Intel directed its customers to write Intel on the side of their products as a means of cobranding. Microsoft liked the idea so much that it now does it also.
6. To become a value-based solution provider. A company switches from selling skill-based products for a fee and a modest profit to providing a value-based product for a percentage of the enrichment the final product gives to the customer. IBM bundled a bunch of the-shelf components made by others into IBM's first PC. The product had a value so much higher than the sum of the cost of its components that IBM unintentionally launched the PC-clone market.

Endorsing this tends towards virtual enterprise, said Cyrus Freidheim, Vice Chairman of Booz, Alien and Hamilton, a firm of management consultants, in the annual meeting of the World Economic Forum at Davos that the current economic and political developments would mean that today's global firms will be superseded by the relationship enterprises. He further remarked that these enterprises will be major corporate organizations with total revenues approaching U.S. $ 1 trillion by early next century, i.e. larger than all but the world's six biggest economies.

These global changes in manufacturing are impinging on Indian economy in general and manufacturing sector in particular. As we step into the citadel of 21st century, we can derive inspiration from the economic progress made in the 1990s. During the year 1998-99, India notched a growth of 5.8 per cent in GDP, which is the second highest in the world. The domestic manufacturing industry could brave the temporary slow down in the capital goods market and is in all readiness to meet the challenge of change. In the Machine Tool Industry, Machine Tool Manufacturers are joining hands to enhance symbiotic relationship amongst them to counter the global onslaught and impart competitive edge to the domestic industry.

The new millennium will be replete with opportunities to be harnessed leveraging the strengths. India's low cost manufacturing base, better systems, liberalised environment, skilled manpower, and growing domestic market for consumer durables, automotive components and accessories, castings and forgings would augur well for the domestic industry. Corporate strategies, such as alliances, better supply chain management with ancillarisation, will enable the Indian industry to carve a niche in the global market. India has the requisite credentials to be an outsourcing base for global majors. The onus now rests on the Indian industry to nurture the intellectual capital with strong industry—academia interface.

The Indian scriptures have also emphasised the need for synergistic efforts for common good or *'Shreyas'*. It is worthwhile recalling here the concept of *vasudhaiva kutumbakam* or the universal brotherhood, as depicted in *Narayanopanishad* and the *Rigveda* hymn which succinctly depicts the concept of togetherness for the prosperity and well being of the universal humanity thus:

Meet together, Talk together.
May our minds comprehend alike,
Common be our actions and achievements,
Common be our thoughts and intentions,
Common be the wishes of our hearts.
So, there may be thorough union among us.

4

Future of Telecommunication and Electronic Equipment Manufacturing

*Air Cmde S.S. Motial**

The advent of scientific management saw manufacturing evolve into an organised process. The advances in technology and resulting mass production brought a system of quality standards, delivery schedules, and cost advantage through economy of scale.

In world class manufacturing today, owing to the essential focus on customizing, quality, delivery, cost and flexibility are the areas in focus.

A holistic view should include proper weightage for market, design and lifetime support to products, individual's knowledge and the environment.

Important events influencing management today are TQM, Business process re-engineering, computers, information revolution, globalization, WTO, etc.

Important enablers to world class manufacturing in the next millennium are Infotech, ERP, convergence of computers, communication, broadcasting, multimedia, automation, robotics, etc. Growth and the ascendancy of the service sector will also have an impact.

How goods are traded and how financial and human capital are to be employed in manufacturing depends on the events and enablers.

Electronic technology has been advancing rapidly to provide customers with products that are portable and handy with advanced features. The consequent miniaturisation in design calls for highly automated manufacturing.

Telecommunication equipment of the future will provide new features/customised solutions through software modifications.

Developing countries like India, where manufacturing is manpower-intensive, face a challenge of bringing about a complementary relationship between technology and human beings to be competitive.

Government policy can make manufacturing a profit-centre and also influence location and logistic decisions.

Strategy today is not a formal plan drawn by professionals. It is founded on shared vision. Top management commitment and a vision for manufacture is essential. An agile organization structure will be the key to manufacturing in the future.

**Chairman and Managing Director, Indian Telephone Industries, Bangalore, India.*

5

Evolution of Technology in Electronics Industry and its Impact on Manufacturing Strategies

*H. Ramakrishna**

Electronics industry has undergone major technological advances during the last fifty years and is expected to continue with the same momentum in the next millennium. These technological advances have had profound impact on the manufacturing strategies of the companies. Companies that were unable to adopt these changes often faced major problems, many a time resulting in closure. This paper attempts to trace the history of the manufacturing industry and to understand the evolution of manufacturing industry in electronics in the new millennium. The ages are classified as the 'Fabrication Engineering Age', 'Assembly Engineering Age', 'Software Engineering Age', and 'Virtual Manufacturing Age'. The essential features of these eras in manufacturing industry and their impact on capital/labour intensity is explained.

PREAMBLE

Electronics industry has been absorbing technological advances which are occurring continuously without let-up in the past fifty years. These technological advances have had tremendous impact on the manufacturing strategies of the electronics companies. If we trace the history of the electronics industry, it can be divided into four phases:

1. The fabrication era,
2. The assembly engineering era,
3. The software product era, and
4. The virtual manufacturing era.

THE AGE OF FABRICATION ENGINEERING

This era which coincided with vacuum tube as the primary electronic device roughly ended in mid-sixties. The product architecture essentially consisted of a few vacuum tubes and associated components like resistors, capacitors, coils and transformers. However, complex products had a large number of mechanical parts like gear trains, springs, cams, rotary dials, knobs, tuning capacitors and many types of technologies like castings, forging, sintered parts, plastic parts, variety of plating, painting finishes, and variety of sheet metal parts. Truly, it was a mechanical engineers' delight and the industries were termed as light engineering industry. Tool designs had dominant role in improving productivity and the role of electronics engineers was relegated

**Chief Scientist, Central Research Laboratories, Bharat Electronics, Bangalore, India.*

to the fag end of a production line, where he performed a few tests to ensure that the products were doing the intended function. Assembly was heavily labour-intensive and many industries took the licensed production route bringing the cost-advantage of low-cost labour in assembly-intensive operations. Capital costs were incurred mainly in the fabrication set-up and test equipment required in final stages. There was considerable scope for components manufacture wherever economics of scale existed and some developing countries, including India, were able to achieve 100 per cent indigenisation with the help of government policies which protected the component industry. This era also saw the sprouting of ancillary units for various large industries, as well as small scale industry movement. With low capital requirements and government incentives, particularly for starting industries in backward areas, some of these units thrived and grew substantially in this era.

THE AGE OF ASSEMBLY ENGINEERING

With the advent of solid state devices, integrated circuits and printed circuit boards, the content of mechanical parts declined rapidly. The Spartan look of the assembly shops with the ubiquitous soldering iron and tweezers and nose pliers yielded place to a mind-boggling variety of automated assembly shops with vibrating feeds; components lead cutters, benders and automatic insertion machines and wave-soldering equipment. Capital costs in setting up assembly plants exponentially increased resulting in small scale sectors virtually folding up. Surface mount components and assembly plants, using robotics, brought down costs of production further. Japan leveraged these techniques in the consumer electronics area virtually edging out U.S.A., whose initial product launches were emulated by Japan with drastic cost reduction using novel manufacturing strategies. The techniques pioneered by Henry Ford stood the test of time in this era with complex tasks being divided into small discrete tasks assigned to each person who would do repetitive jobs with monotony of a robot. This practice was used in many developing countries who could not afford the cost of highly automated assembly set ups. At about the same time, the reduction of mechanical parts and module-based product designs lent itself to automatic testing. The ubiquitous computer, starting with Apple computer and then the IBM Personal Computer found its way to production floors as CNC controllers, as automated assembly and inspection equipment. But the major revolution was in the automatic tester which became a dominant technology. The Hewlett-Packard pioneered HPIB bus (later IEE 488) which, with simple English language like commands, evolved a breed of test engineers, who could write programs which could reduce test cycles, remove testing bottlenecks and exponentially improve shop floor productivity.

THE AGE OF SOFTWARE PRODUCTS

The software, which was once confined to the backrooms of EDP for pay roll and financial computerisation, started playing a predominant role in manufacturing industries in the eighties. Apart from a host of software products for manufacturing resource planning, project scheduling, which gave substantial productivity benefits, embedded software became a dominant part of any modern product architecture. It appeared in the form of Application Specific Integrated Circuits (ASIC), which are designed using sophisticated CAD tools. The versatile microcontroller in every conceivable product allowed product customisation down the line for specific customer segment with relatively simple programming techniques. In the nineties, the advent of the digital signal processors (DSP) ensured a predominant place for software in all types of products.

This had a profound impact on countries which had thrived on cost reduction strategies of products designed elsewhere. The dominant role is now back in the hands of original conceivers who can give effect to their nascent ideas on general purpose complex chips. The complex chips, in any case, are manufactured only by giants like Intel, Motorola, Texas Instruments, etc. Components industry of any sophisticated variety is no longer viable in any developing country. The process of licensed production is becoming irrelevant as not much value can be added locally, and, in any case, in an open economy, no major manufacturer likes to disperse his manufacturing activity at various geographic locations. Reverse engineering which was considered as merely copying, but, effective process for developing countries, is no longer valid as most of the product architectures resides in embedded chips which are very hard to copy. Thus, developing countries have taken the software export route for growth and, in future, can make significant contribution in domestic products as well.

THE AGE OF VIRTUAL MANUFACTURING

The nineties have seen a revolutionary change in the way we do business. The traditional manufacturing, as we were used to, is undergoing change. The brain of electronic products which used to be a processor, and a number of peripheral chips are continuing in their relentless path of high complexity and low cost. We now hear of "system on a chip" products. The assembly has become "impersonal" in the sense, low cost assembly can only be done by high capital cost outfits who addresses production on a grand scale with attendant economics of scale and agile business operations. A product concept goes through a virtual route before it reaches the intended customer. The design itself will heavily depend on the purchases of a set of "IP" cores, which are virtually assembled and simulated to achieve the required function. The so-called design with the list of "IP" cores, which have to be licensed, can be taken by a business house who can get the VLSI fabricated in any foundry and reserve the right to market it. Another business house may purchase such complex components and realise a product by customizing with added software and get printed circuit assemblies fabricated in contract manufacturing outfits and release product kits. There may be other business house buying such kits and add the required housings and further customise for identified markets and sell to the end customers. With the advent of languages like 'JAVA', high level of customisation can be done independent of the platform used. Thus, the traditional concept of manufacturing a product using manufactured and purchase parts under a single roof and assembly and testing to release a product is rapidly vanishing. It would, therefore, be difficult to say who is the real manufacturer. However, depending on the skill set and core competency of a business house, he will try to capitalize on that portion of manufacturing which suits his organisation. The concept of delivery chain will not have visibility of every one effecting the transition making all parts of the process except the one you are involved and seem virtual to you. Developing countries like India who have manpower resources capable of lateral thinking and software programming would do well to concentrate on the mass customisation segment at one end and development of "IP" cores at the other end. The middle portion of the manufacturing, which seem real to us, that is making semiconductor chips of high complexity and printed circuit board assemblies with highly automated assembly lines are slowly becoming beyond reach for traditional manufacturers, because of the high cost of the setting up and agile operations required to deliver low cost product. In the new organisation, "the task workers" will look like some relic of old industrial age. The worker will become an intelligent part of overall process.

"Mass Customisation" which combines the efficiencies of high volume production with ability to build customer specific products speedily and then the journey through a virtual manufacturing road will begin till the customers actually get products in their hands.

Electronics industries have been changing rapidly over the years because of the rapid strides made in the enabling technologies. The broad phrases have been the fabrication era, the assembly engineering era, the software-based product era and is on the threshold of the new millennium heralded by internet-based business strategies to an era of virtual manufacturing. There is scope for developing countries to make significant contributions by choosing the segment appropriate to their skill set.

Future of Manufacturing Technology

*A.V. Hanumantha Rao**

Undertaking product manufacture with proven manufacturing technology, popularly known as "know-how", is very advantageous, but for non-commercial reasons import of know-how may not always be possible. With few exceptions like a small two-stage sounding rocket, way back in 1960s, pressure transducers and liquid engines, Indian Space Research Organisation, Department of Space did not import any manufacturing technology to realise products for any of its programmes. Almost all of the manufacturing technology required for the launch vehicle hardware and material is home grown, thanks to the enterprising Indian industry. Of course, dependence on external sources remains in creating facilities, towards procurement of special category of machine tools. But due to a variety of reasons, procurement of some facilities is not always possible. Hence, some of the manufacturing technologies adopted by ISRO are inferior to the ones developed and practised by the advanced countries. In this situation, the probable thrust for manufacturing technology in the future, in respect of ISRO, would be to strive to develop the technology, along with the relevant machine tools in the country, so as to be at par with advanced countries. Typical examples are large ring forgings and flow forming of large cylindrical casings for launch vehicle programmes of ISRO.

INTRODUCTION

Axi-symmetric and circular cross section is the typical configuration of a launch vehicle. This facilitates application of standard machine tools for manufacture of component parts and assemblies. For maximising payload capability of the vehicle the designs are highly optimised to achieve least inertia weight. The design margins are kept very low, only of the order of 1.1. This demands strict control on processes and manufacturing technology at every stage, right from raw material to finished product. This also dictates the choice of materials. In a multi-stage rocket the inertia weight of the lower stages has lesser influence on payload sensitivity compared to the upper stages. The lower stages are generally constructed out of metallic materials. Almost all types of manufacturing processes like forming, rolling, welding, machining, etc. are employed in realising the hardware.

PRESENT MANUFACTURING SCENARIO OF MAJOR HARDWARE

The hardware for the solid propulsion stage of the launch vehicle is essentially a pressure vessel. This is designed to carry propellant and also serve as a combustion chamber. Thus, this hardware has to withstand the internal pressure buildup during the ignition and combustion

*General Manager, Mechanical Engineering Entity, Vikram Sarabhai Space Centre, Trivandrum, India.

of the propellant and expansion of the products of combustion through the nozzle. Also the inertial and flight loads are to be considered. The design principles are predominantly the same as of a pressure vessel. The casing is designed to realise the hardware by rolling and welding of plates to ring forgings. But the demand of least inertia weight with low design margins makes the process of welding highly demanding. Some of the high strength materials, used for the construction of the pressure vessel, are prone to fracture. Here the design has to be fracture-proof. The fracture toughness of the weld material is lower than the parent material. The plates used to form the cylindrical shells also do exhibit waviness and consequently the joint quality is prone to weld mismatches. Some of the weld joints also undergo multiple repairs. These limitations of the technology do not permit full potential of the material to be exploited by the design.

SUPERIOR MANUFACTURING TECHNOLOGY

To overcome the above limitations, weldless construction is resorted to for realisation of the cylindrical segments of the pressure vessel. The process is known as flow forming, which is akin to a potter giving shape to a mound of clay on potter's wheel (Fig. 6.1). Equipment for flow forming the material on a mandrel in lathe type of machine is available in the country and is being used for smaller casings. For larger casings, having an overall dimensions of 2800 mm diameter, wall thickness of about 6.5 mm and 3500 mm of length, there is no facility for flow forming. The input material form for flow forming is a ring forging, having the same mean diameter and volume as that of the end product.

FLOW FORMING OF LARGE CYLINDRICAL CASINGS

The architecture of the machine is conceptually identical to a potter's wheel (Fig. 6.1) but with one difference, *viz.*, multiple number of potters work simultaneously, on the same material and wheel, as depicted in Fig. 6.1.

The process sequence of flow forming is shown in Fig. 6.2. The ring blank fastened to the annular table of the machine is made to rotate about its axis. Four pairs of rollers, staggered axially relative to each other, engage into the inner and outer recess pre-machined at the top end of the blank. The forming rollers move radially in and out of their respective housing to control the product thickness. The inner and outer carriages, which carry the roller housings, traverse vertically down causing progressive reduction in wall thickness to the desired length.

Wall thickness reduction, as high as 80 per cent to 85 per cent have been achieved. Compared to mandrel type of machine, the forces on the rollers are halved and also the end product has very insignificant residual stresses. This technology of chipless manufacture yields geometrically and dimensionally accurate and stable product with improved fracture toughness and other mechanical properties. Typical tolerances obtained on a large 3 m diameter casing are shown below:

Diameter	+0.08%
Wall thickness	+1.2%
Circularity	1.0 mm
Perpendicularity	1.0 mm

Cost benefit and economic viability of this technology have been realized by ISRO.

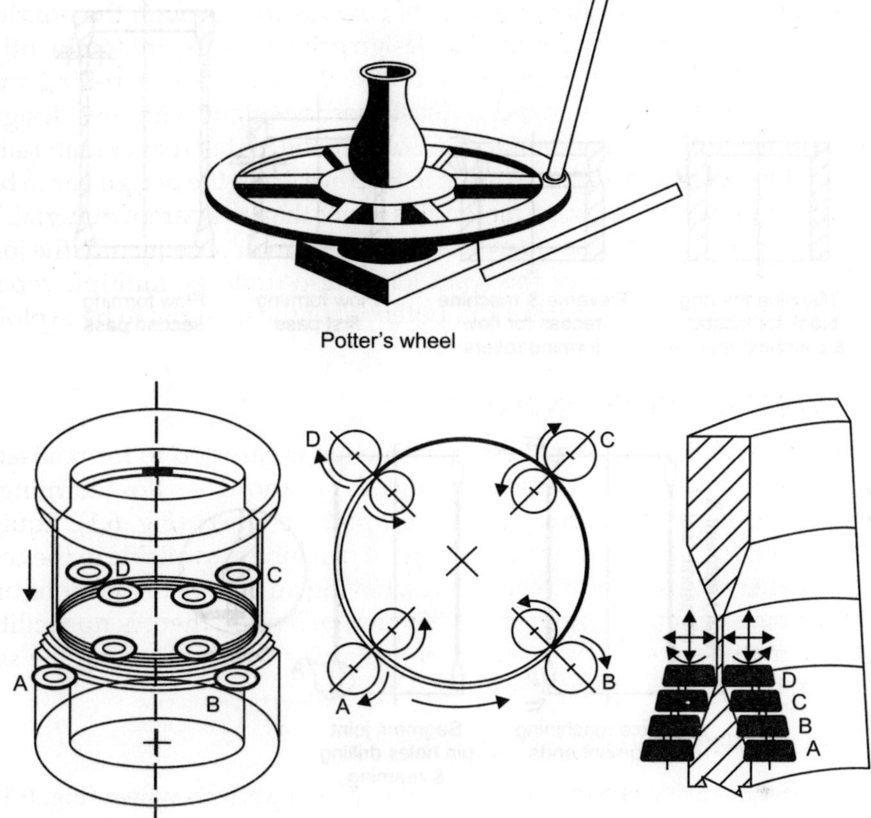

Fig. 6.1. **Flow forming concept: Roller pairs superimposed in one plane to show the relative difference in elevation and radial gap between them**

RING FORGINGS AND PROCESS

Ring forgings are widely used in almost all the structures of the launch vehicles. Ring forgings out of a variety of materials, *viz.*, aluminium alloys, titanium alloys, high strength steels, super-alloys, etc., are required. Manufacture of these forgings requires a ring-rolling mill. Of the total requirement of ISRO, about 30 per cent fall outside the capacity of the mills existing in the country. Absence of a mill of required capacity has led to engineering of products as welded/ riveted assemblies of multiple piece parts. The input material for flow forming is also a ring forging. Some of ring forgings could be realised by mandrel forging. But mandrel forging consumes excess raw material and energy. It also suffers from non-uniformity of grain flow and mechanical properties. Hence, it is not a preferred process for critical applications. Realization of product from a shaped and/or integral ring forgings makes a significant reduction in the product manufacturing operations as well as improved quality. Typical example of a launch vehicle product (nozzle forging), which has vastly improved the product quality and reduced the cost, is shown in Fig. 6.3.

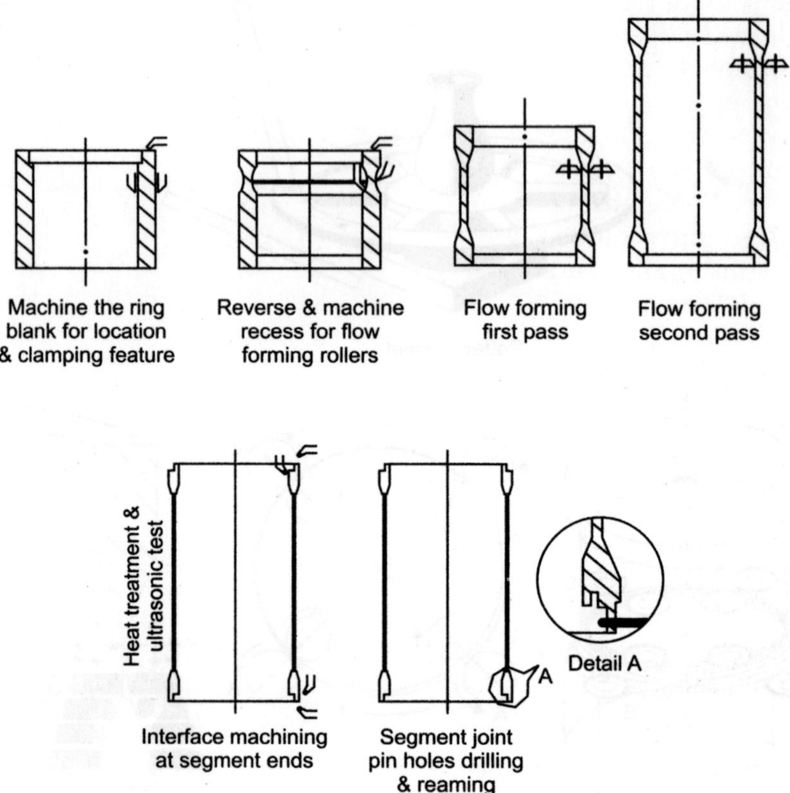

Fig. 6.2. Flow formed segment realisation from ring forging-process sequence

RING ROLLING MILL

Conceptual layout of a typical ring rolling mill is shown in Fig. 6.4.

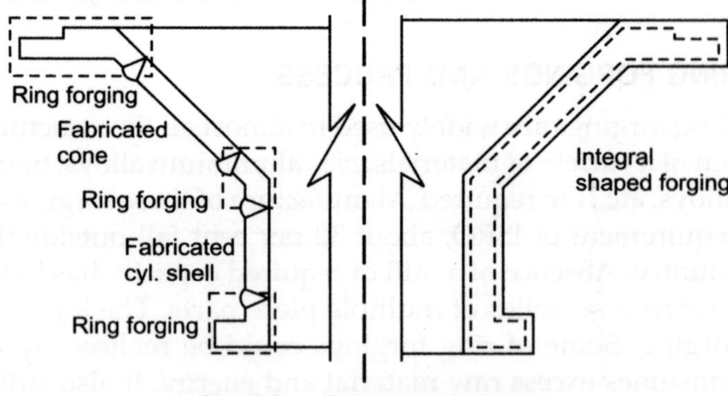

Fig. 6.3. Typical nozzle forging

The hot metal in the form of pancake, having a pierced centre hole, is loaded on the mill and the mandrel is inserted into the hole. The radial thickness of the pancake gets progressively squeezed between the idle mandrel and power driven main roll, causing growth in diameter and height of the ring being rolled. The centring rolls and axial rolls control the circularity and height respectively, of the ring being rolled. Shape given on the mandrel or the main roll imparts shape to the respective side of the forging.

Typical process sequence for ring forging manufacture is shown in Fig. 6.5.

In essence, the ring rolling process follows the same sequence as that of the mandrel forging. The principal difference is mechanisation and automation and in-process control.

Setting up a facility of the required capability should lead to emergence of improved designs in other engineering applications as well, which may improve the low utility of the mill for the limited requirements of ISRO. Number of machines, on the entire globe, which can meet the higher end requirements of ISRO are not more than four or five. The price commanded by mills of this class varies from 2.25 times to 5 times per ton weight of large and sophisticated multi axis CNC machine tool.

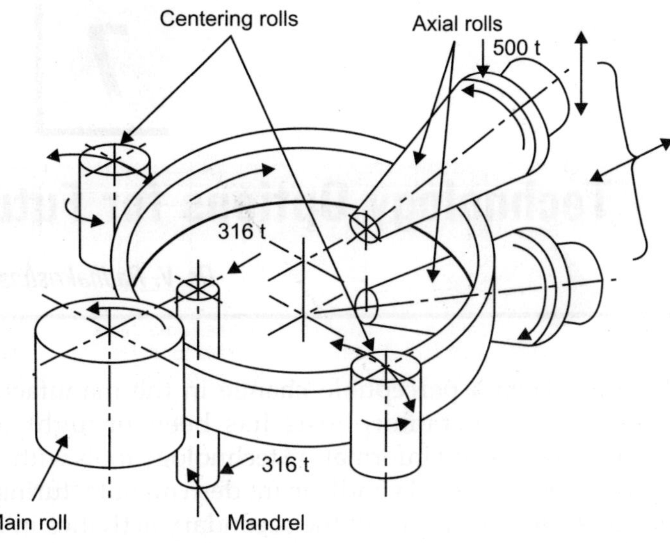

Fig. 6.4. Principle of operation of (radial-axial) ring rolling mill

The highly optimised designs and quality demands of ISRO are expected to provide some challenging motivation for the manufacturing technology in the future. The setting up of and availability of facilities of the above class is expected to improve the product designs in other engineering fields as well.

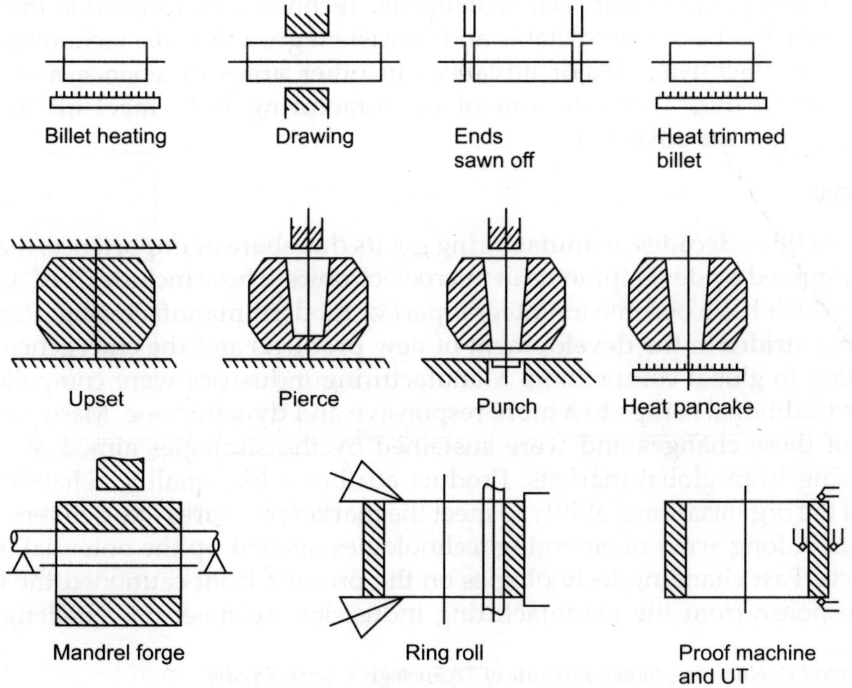

Fig. 6.5. Ring forging realisation process sequence

7

Technology Options for Future Manufacturing

*Dr. V. Radhakrishnan**

There has been a perceptible change in the manufacturing sector in the past two to three decades. This metamorphosis has been brought about by the timely integration of microelectronics and information technology tools with shop floor manufacturing. The synergy of this can be witnessed vividly in modern manufacturing. Manufacturing technologies adopted and the system covering all the secondary activities in manufacturing has been benefitted by this. There were even predictions of unmanned factories and computer integrated manufacturing monoliths, which have not come through due to pragmatic considerations. However, the application of new technologies and system tools have remarkably enhanced the quality and productivity, in the recent past. Technology is highly independent of the manufacturing environment whereas the system part is governed greatly by the physical environment prevailing within the industry and surroundings. Thus, optimising the system often goes beyond the four walls of the factory, putting certain constraints on the benefits, which can be derived. As against this, technology options are self-sustainable. Technologies adopted in manufacturing are not static. There has been a predictable and consistent growth in the technologies adopted for advanced manufacturing. Major advances in other areas of science and technology have contributed to this. Ultimate aim of manufacturing is to meet all the customer demands—whoever is the customer.

INTRODUCTION

In the past two to three decades, manufacturing got its due share of importance, thanks to new technologies triggered by developments in microelectronics. These include CNC, CAD/CAM, Robotics, etc., which have become an integral part of modern manufacturing. Parallel to this there were rapid strides in the development of new products and the emergence of an open economy leading to global competition. Manufacturing industries were compelled to move away from the traditional setups to a more responsive and dynamic one. Many new concepts emerged out of these changes and were sustained by the strategies aimed at meeting the challenges arising from global markets. Product attributes like quality, reliability, cost, life prediction and the organisational ability to meet the market pressures like delivery and service came into focus. A long array of emerging technologies opened up the potential for a variety of new products. Fast changing technologies on the product front cautioned the need for an equally fast response from the manufacturing industries. To meet this challenge, with the

**Professor, Mechanical Engineering, Indian Institute of Technology, Chennai, India.*

assistance of advanced manufacturing technologies, industries moved from rigid systems to flexible and cellular concepts. Organisational changes, effective information integration and business process re-engineering made it possible for many industries to be lean and agile. Conceptual changes in product development like concurrent engineering showed visible product lead-time reductions. Application of well-established software for enterprise level resource planning started giving optimal results with reference to organisational aspects in manufacturing. The results were dramatic. In organisations the mess up was maximum. This change in thinking allowed top management to re-look at factors, which were earlier considered as insignificant. Their significance came to light with improved profits and other tangible benefits accrued by the new system. These included a significant reduction in inventory and work in progress, predictable schedules and better money management.

MANUFACTURING SYSTEM

Response of an industry to the market demand depends, to a great extent, on the manufacturing system adopted by the organisation. Optimising the system performance with reference to any given objective gives the best results in a competitive environment. However, any optimising procedure is as good as the significance and reliability of the available data. Hence, there is bound to be some difference among industries implementing the same system concept. One with a better understanding of the influencing parameters and with integrity and reliability in information will fare far more better than others lacking in these qualities. By and large, **system optimisation** beyond a level is not cost effective. Many organisations neglect the significance of manufacturing technology in global competition. To be competitive, in addition to system optimisation, it is essential to have **technology optimisation**. This comes from the technologies adopted in manufacturing covering tooling and processing. Saving in processing cost is negligible in single piece manufacturing. The significance of this is felt when product variety and batch sizes increase and in mass production.

MANUFACTURING TECHNOLOGY

The processing and tooling requirements connected with manufacturing are discussed here. The technologies adopted for product realisation are equally critical and important as the system adopted in realising quality products is available at competitive prices. Though the principle of working of a product is well known, it is the processing technologies involved in its realisation, which is tough to master. Almost 75 per cent of the production costs are decided at the time of design. This includes material, design details, component geometry, positioning, tolerancing, finish, etc. They, in turn, decide the processes and process parameters. If any process modification or substitution is to be done, certain design changes are essential. Examples could be seen in the replacement of machined parts with cast or forged parts or in the substitution of metal parts by plastics. Many of these procedures add value to the product and reduce the cost substantially. There are many operations during manufacture, which are not directly connected with the shaping of the part; often these are done off-line. These include heat treatment, part lubrication, part separation, cleaning, etc. If these processes are integrated with the production line operations, they effect considerable savings. With advances in manufacturing technologies, it is possible to reduce the process chain normally required for the realisation of a product. Assembly operations are considered to be time consuming and labour intensive. By adopting suitable manufacturing procedures it is possible to save time and re-work in assembly.

There is a need to develop process technologies that are environmentally friendly. In brief, the area of manufacturing technology is of critical importance to future manufacturing industries. Adopting the right technologies can significantly improve the profit margins and keep the competition away. This is a continuous process and innovative concepts are essential for continued success.

PROCESS CHAIN REDUCTION (Fig. 7.1)

Design requirements, material, product geometry, etc., normally decide the processing sequence adopted for the realisation of the product. Over the years, time tested processing sequences have been adopted without looking into other alternative sequencing of the operations. In certain cases, it is possible to eliminate some processes without affecting the product quality or reliability. Savings in time and tooling are often great and can substantially improve the competitive advantage. In many cases, this approach may add value to the product through performance improvement of the part. Cost savings are not only effected by the elimination of certain operations and associated machinery and tooling but also by the elimination of processing defects. Presently, there are many technologies that are available to achieve this goal. An example is **hard turning**. Here a costly process like grinding is eliminated and the rejections due to hardening done between operations are almost absent. **High speed machining** makes the operation fast thus reducing the cycle time, improving the finish and producing the workpiece free from surface damage due to heating. Chip disposal also becomes easier in some cases. Today, both hard turning and high speed machining are possible with improved design of machine tools and the availability of new cutting tools. The new tool materials used include CBN, coated carbide tools and diamond.

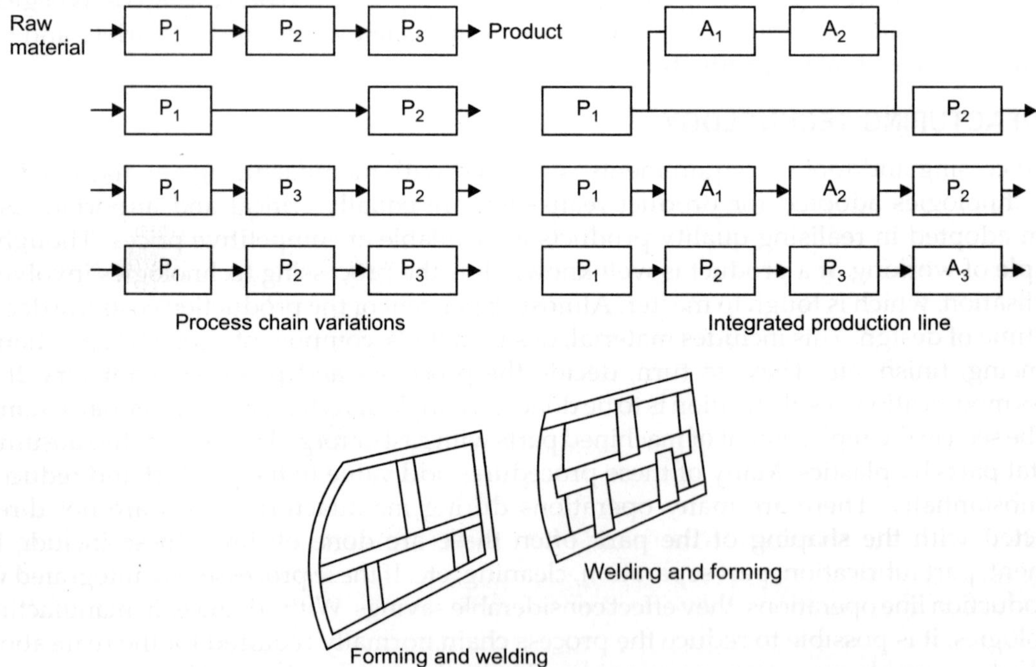

Fig. 7.1. Schematics of process chain reduction and integration with an example

There are many possibilities to substitute a process by another one, which may give a better quality and reduced cycle time. **Net shape manufacturing**, can cut down the process chain dramatically. In certain cases, no machining may be required and in other cases finish machining can be done directly. Cold forming is an ideal net shaping process. Roller burnishing, sizing and finishing can improve the surface strength, tolerance adherence and finish requirements on the part. Often, this is an integrated process, which reduces the cycle time considerably.

Another process tried out is internal **high pressure forming** which eliminates the need for punches and two sets of dies and presses. **Hard roller burnishing** can replace honing in certain applications. It is essential to study the product requirements and look at the sequence adopted presently to evolve a strategy for modifying the sequence. This should also take into consideration any possible material substitution.

VALUE ADDITION THROUGH NEW TECHNOLOGIES

Materials and parts are processed to give them the required design characteristics. The design is done taking into account the performance requirements of the product. By adopting a new process or modifying an existing one substantial value addition can be given for the product. A tin coating can substitute gold coating giving both the appearance and better coating hardness for the product. Coated steel sheets are value-added products and by developing the coating technology so that the sheets can be formed to give considerable flexibility and quality to the product produced from these sheets. **Laser treated steel** for automobile sheet metal parts gives better formability and surface characteristics due to better lubricant retention (Fig. 7.2). Surface technologies covering coating, microalloying, selective surface hardening, glazing, etc. provide process time reduction and value addition.

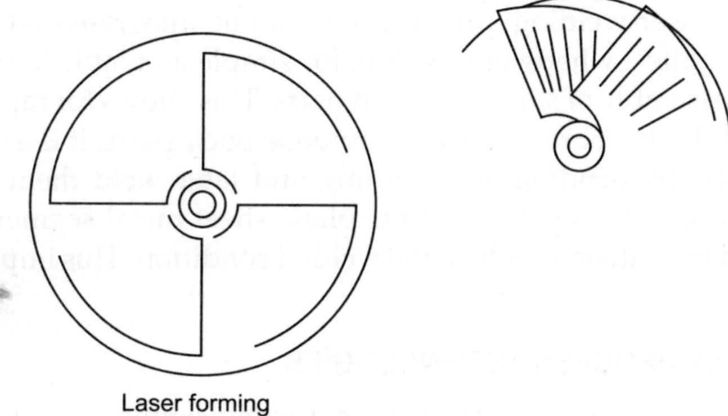

Fig. 7.2. Laser forming of sheet metal

PRODUCTION LINE INTEGRATION OF OFF-LINE PROCESSES

In manufacturing, there are many operations, which are not directly connected with the final shaping of the part. These include workpiece preparation before the process like melting, heating, lubrication, cleaning, etc. In many cases, the workpiece is to be separated, positioned, stapled or magazined before being fed to the process. This is true in assembly operations. In between heat treatment means that the parts are to be parallel processed. This is the case with deburring and surface treatments like galvanising, coating, and painting. Most of these operations are difficult to be integrated in the production line due to the long processing time needed for them. If integrated, on-line without proper planning, costly equipment on-line will be put to inefficient use. Often, these operations are done off-line through by-pass lines making them a costly proposition. In turn, this increases the product cycle time. In many cases, the production time is of the order of a few seconds but the preparation time may be few hours or days. Any success in integrating such processes with the production line gives excellent rewards. As an example, one can mention the introduction of surface hardening using induction hardening replacing through hardening done off-line. UV heating of paints and integration of measurement (on line) and inspection procedures. Selective hardening using laser and *in situ* hardening using the heat generated from grinding. Technological advances in inspection have provided a vast scope for such integration of inspection systems. **Laser based inspection** devices are fast, reliable and amenable for such integration. **Computer vision application to inspection** is proving cost-effective and ideal for on-line integration.

MANUFACTURING TECHNOLOGY AND ASSEMBLY

Assembly is the culmination of all production processes. Design for assembly is done at the design stage taking into account the assembly requirements, ease of assembly and inspection. Often, the production facilities are not considered in parallel with the assembly procedure to be adopted. This results in the non-integration of parts during design, leading to more parts and more problems in assembly. Present technologies allow the production of complex parts with ease, and their assembly is also simple. Machining centres, turn-mill centres, laser integrated sheet metal turret punch presses are examples in the production of such complex parts. In certain cases complex tools can be used to achieve this objective. Modern assembly techniques like gluing, snap joining, laser welding, pressing, etc. can be integrated with the production line to produce sub-assemblies, wherever possible, for simple assembly later. Final assembly should be planned as an assembly of sub-assembled parts. This allows for rapid model changes in flexible assembly lines. In the fabrication of automobile body parts, it is a common practice, to make sheet metal parts by bending and forming and later weld them together using a fixture. A fast and better approach is to weld the plane sheet metal segments first with low positioning accuracy and form them finally in the welded condition. This improves quality and speeds up the process.

ENVIRONMENT FRIENDLY PROCESS TECHNOLOGIES

The coming decades will witness increased sensitivity for environmental aspects and there will be strict laws regarding environment. Manufacturing processes should eventually become environmentally friendly. If less material is used in manufacturing, it can reduce the energy consumption and the waste. There are many areas in manufacturing which need attention from the environmental angle. These include coolants and oils, heat treatment and electroplating, cleaning fluids, painting and coating, chips and swarf. Presently, there is no technology available to eliminate the use of these auxiliary materials completely from the manufacturing environment. The thrust is on the improvement of the present conditions before the technologies are developed to eliminate such environmental hazards. This could be done through avoidance of dangerous materials being used in the shop floor (e.g., cyanide salts). Other steps include providing closed cycle treatment for the pollutant material before recycling them, filtering and cleaning of pollutants before releasing them and providing proper workspace enclosures and worker protection in the work place. Efforts should be directed to identify alternate solutions including new compositions. However, the ideal solution is the total removal of such pollutants in the production line.

Dry cutting is aimed towards this goal. There are, however, many factors to be considered with reference to the introduction of this process. These include product quality and production time. In dry cutting, heat generation affects both tool and workpiece, resulting in quality and production problems. Workpiece dimensional control becomes difficult with increase in heat input to the workpiece. Material characteristics could get altered due to microstructure changes and the tool wear may become substantial needing frequent interruptions in the process. Coolants are essential for heat transfer, lubrication and chip disposal. Cost of coolant per finished workpiece is more than the tool cost per piece. This includes the cost of equipment, its maintenance, energy needs, coolant cost, etc. Thus, a reduction or avoidance of coolants could also be economical. Attempts to machine with limited coolant supply have been successful and application of cryogenic cooling is attempted in this direction. Heating the workpiece at the cutting zone by laser and machining it without coolant is again another technology in this direction. Processes like grinding, where the coolant requirements are very high, are to be studied in detail and use of environmentally safe coolants and lubricants, which are biodegradable, are to be investigated. Chip recycling is viable both technically and financially. However, this should be integrated with the plant design to get the best results. Considering the fact that chip production is high in modern manufacturing, any step towards recycling is worth the effort. Product recycling has given a new dimension to manufacturing. In future, both product assembly and disassembly are to be considered. Technologies are to be developed for fast and foolproof disassembly of products. Naturally, this has a lot to do with the manufacturing technologies adopted for the realisation of the product.

COST EFFECTIVE SINGLE PIECE MANUFACTURING

Advanced rapid manufacturing technologies have given the option for producing even a single-part with cost effectiveness. Though primarily these procedures are used for prototype manufacture and for rapid tooling, there is the possibility for some of them to be used directly for product realisation. These include selective laser sintering, laminated object manufacturing using metal sheets and their effective bonding and superplastic forming. One is not sure how soon some of them may form the basis for regular production approaches.

PRECISION ENGINEERED AND MICRO-MACHINED PARTS AND PRODUCTS

Value addition increases as the tolerances and dimensions given on the parts decrease dramatically (Fig. 7.3). Precision manufacturing and micro machining are two closely related technologies, which are having a significant impact on a variety of new products being developed. The materials used and the design concepts involved are different in this area when compared with conventional manufacturing. Technology for the manufacture and inspection of such parts and products are difficult to master. This needs a fundamental understanding of the technologies suited for manufacture and practical experience in such manufacture. The

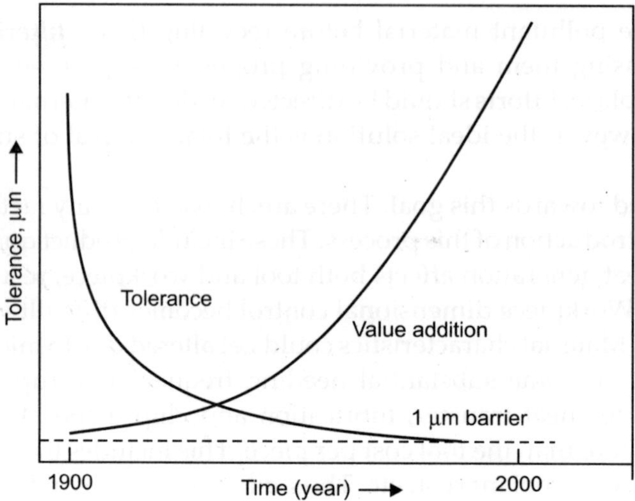

Fig. 7.3. Value addition in precision engineering

demand for such products is on the increase and more products are coming under this grouping. Organisations that have mastered these technologies will profit immensely from these developments. In the production of micron and sub-micron parts, the business opportunities are immense.

NEW CONCEPTS IN MACHINE TOOL DESIGN

Machine tools are considered as the backbone of manufacturing industries. Here the discussion is confined to the application of new machine tool concepts in enhancing the manufacturing productivity and competitiveness and not on the advances taking place in the Machine Tool Design, as such. A major thrust in the development of advanced machine tools has been to reduce the machining time, idle time, handling time and setting time. Examples include two spindle machines, which can machine on both sides of the part. In a top and bottom spindle configuration, the bottom spindle can also provide the transport of the part to the next station. There have been new concepts regarding the design of interchangeable pallets for spindle mounting. They carry coded information on the part fixed to them and can be transported fast between machines (4 m/sec). High spindle speeds are provided for machines to decrease the machining time as well as to improve the machining quality. Lower cutting forces encountered at higher speed permits the machining of thin walled parts without any distortion. High-speed milling eliminates chatter due to cutting frequency much higher than the critical value. Burr formation is almost absent in high speed machining. Spindle speeds above 25000 rpm are already available in machine tools and machine tools with linear motor slides have also been developed. Vertical pallets are found to be better as the chips do not settle on them giving thermal problems in high speed machining. To reduce setup time, dynamic feed drives are adopted in certain machine tools. In multi-cylinder engine crankshaft grinding more than two settings are needed and there is the difficulty in maintaining the accuracy due to unbalanced weights. A new grinding machine concept with the wheel head moving on a secondary slide for plunge movements can reduce the setting and idle time considerably (Fig. 7.4). Fixturing is simplified and the machine is flexible to handle any crankshaft. Currently, most of the machine

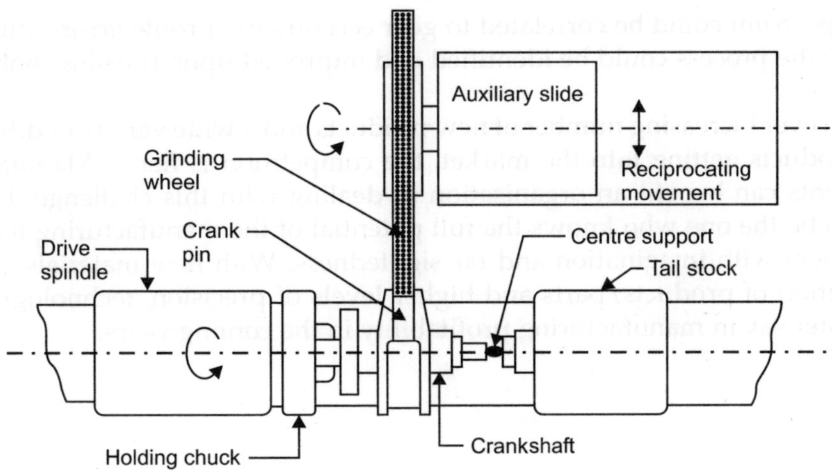

Fig. 7.4. New concept of a crankshaft grinder

tools use open kinematic chain in their design. Here each axis must move or carry other axes positioned further along the chain. To overcome this defect, machine tools are designed with complex kinematic structures. These are known as parallel kinematic structures. Advantages of this design include reduced mass of moving parts, increased acceleration, modular construction and high stiffness. Example of this design is the Hexapod. Hexapod design of machine tool kinematics gives better stiffness and can accommodate five axis capabilities. Many of these innovations in the machine tool area are to be critically examined for their adaptability to the product being manufactured to get the best results.

ADVANCES IN INSPECTION

As quality requirements are on the rise, there is bound to be more quality-related activities to be performed in manufacturing. However, cost reduction is essential in inspection and time taken for inspection should be minimum. This means that inspection should be speeded up, should be integrated with manufacturing to reduce time and the introduction of effective process control strategies. Optical and vision system inspection is replacing conventional approaches in the shop floor. CMMs are used directly in the shop floor for inspection. New measurement technologies involving miniature quartz probes permit the inspection of thin parts with extremely low measuring force. All visual inspections are replaced by simple machine vision setups. Grid projection approaches cover large forms of inspections with ease. Process control is achieved through automated measurement strategies, data processing, statistical evaluation, interpretation and automatic process control. Tool wear and machine tool condition monitoring are done to ensure process stability. In-process monitoring of parts is being tried out leading to automated process control loop.

Better understanding of QFD, through product performance evaluation and process error identification, has improved the process technologies to meet the customer requirements. For example, the noise level of the automobile gearbox can be tested in an anechoic chamber and the signal spectrum can be analysed. Specific frequencies observed could be correlated to the component geometry error leading to process defect identification. In the case of the gearbox, the noise spectrum could be correlated to gear eccentricity, profile errors, pitch error, finish, etc. In turn, the process could be identified and improved upon (turning, hobbing, grinding, honing).

With an ever increasing number of new products and a wide variety of different models of existing products getting into the market, the competition is fierce. Manufacturing system improvements can benefit an organisation in dealing with this challenge. But the ultimate winner will be the one who knows the full potential of the manufacturing technologies well and uses them with imagination and far sightedness. With new materials, product ranges, miniaturisation of products/parts and higher levels of precision, technology will definitely have a greater say in manufacturing profitability in the coming years.

8

Pattern of Organization and Manufacturing Management

*Dr. M. Adithan**

MANUFACTURING TECHNOLOGY

Manufacturing has an important place in the economies of all the countries, whether developed, under-developed or developing. Manufacturing plays an important role in terms of employment generation and in its contribution to wealth generation. The advancement in manufacturing technologies in the recent past has brought about a metamorphism in the world industrial scene.

The rapid growth in the world wide application of microelectronics and computers is providing new opportunities for manufacturing processes, products and manufacturing control systems. It is also increasing competitive pressures. Developments in basic sciences and manufacturing processes have also brought a revolution in technology.

Progress in any field could be achieved by multi-disciplinary approach. Manufacturing is multi-disciplinary subject. The impact of computers in manufacturing and in manufacturing related activities is tremendous. We have new technologies, such as CAD (Computer Aided Design), CADD (Computer Aided Design and Drafting), CAE (Computer Aided Engineering) for Analysis, CAPP (Computer Aided Process Planning), CNC (Computer Numerical Control). CAM (Computer Aided Manufacturing), CAI (Computer Aided Inspection), Testing and Quality Control, CIM (Computer Integrated Manufacturing) and CAX (Computer Aided X, anything).

No doubt, investment in the above advanced manufacturing technologies by the companies is costly and needs considerable initial funding. Fortunately, significant trend is taking place in computer technology; computers are becoming smaller, cheaper but powerful and easy access to computing power by more and more people in the organization is possible. The benefits of using these advanced technologies are many, especially for the manufacturing industries.

BENEFITS OF USING ADVANCED MANUFACTURING TECHNOLOGIES (AMT)

1. Reduction in materials costs and total production costs (i.e. Improved manufacturing economy).
2. Higher operating profits (i.e. profitability improves).
3. Reduction in overheads, floor space and operating expenses.
4. New technology allows working capital to be significantly reduced.

Professor, Department of Mechanical Engineering and Dean, VIT, Deemed University, Tamil Nadu, India.

5. Quick changes in design considering customer's feedback, more responsive to changes in market (i.e. companies become more agile).
6. Better Quality Assurance (leading to higher productivity and competitiveness).
7. Above all, greater control over every aspect of production and manufacturing.

A PARADIGM SHIFT: FUTURE MANUFACTURING

Changes that are likely to take place in the manufacturing environment and technology in the first decade of 21st century:

Old Paradigm	New Paradigm
• Inspectors responsible for quality	Workers responsible for quality
• One worker at machine	Self-directed work teams at machines.
• Static job assignments	Worker empowerment and job rotation
• "Management thinks, you do".	"Management and workers think and do".
• Quantity over quality	Quality over quantity
• Price and supply	Quality and customer service.
• Competition	Collaboration/Manufacturer networks
• Individual incentives	Group incentives
• "Let the buyer beware"	External and internal customers
• Local orientation	Global orientation
• Single-job skills	Job clusters/Skill families or Multi-tasking
• Muscle power	Smart machinery
• Machine operator	Knowledge worker
• Individual efforts	Partnerships
• Sporadic training	Constant training
• "Degree" education	Lifelong or Competency based learning
• Multi-tier organization.	Lean organization (Flat organization)

But the question is: How can the manufacturing organizations and their managements take care of the changing scenario and provide for it?

STRATEGIC PLANNING ISSUES

Introduction of new technologies in workplace calls for new pattern of organization and manufacturing management. Organizational pattern and manufacturing management to be successful, effective and efficient in the future will take into consideration the following strategic issues and provide for it:

- Innovation and flexibility,
- Creative climate in the organization,
- Reorganization, re-engineering and workflow redesign,
- Reduced management layers (flat organization in contrast to pyramidal structure), and
- Decentralized structure through autonomous work groups.

- Team based work methods
- Empowerment of people and Total Employee Involvement (TEI) and participation.
- Opportunity management instead of crisis management
- Lean production
- Just-In-Time (JIT)
- Zero-defects
- Co-makership
- Batch size: One, production
- Emphasis on throughput (i.e. improving system productivity instead of individual optimization)
- Continuous Quality Improvement (CQI)
- Continuous Productivity Improvement (CPI)
- Treatment of information as a factor of production
- Investment in education and training in manufacturing technology.

9

Robotics for Hostile Environments

Dr. Andrew A. Goldenberg *

The critical issues involved in the development of robotic systems meant to operate in hostile or hazardous environments are explained here. Several conditions which define such an environment are identified and the design issues relevant for the systems to withstand those conditions are discussed. Case studies of practical developments of such robotic systems are presented here.

INTRODUCTION

Application and relevance of robotics in industry varies from time to time and place to place depending on several factors such as labour cost, batch quantity of production, volume of turnover, etc. However, hostile environments is a field of application that is independent of these variants. It does not matter whether a task is to be performed one time only or repeated, whether the labour is cheap or expensive, or whether the country is Canada or India. An application in a hostile environment almost inevitably needs robotic solutions.

HAZARDOUS ENVIRONMENTAL CONDITIONS

1. **Water and chemicals**
   Robot moves in rain, in marshy fields, underwater, inside chemical liquid containers, or it has to undergo a wash down underwater jets for removing radioactive contamination.
2. **Bad terrain**
   Robot travels on uneven, bumpy roads, rocks, hillocks, gravel, steps, grass, or on sand.
3. **Radiation**
   Robot operates in regions where radioactive emission is higher than permissible for human safety and health.
4. **Fire**
   Robot operates in high temperature areas, such as in furnaces, fire fighting tasks, and metallurgical plants.
5. **Vacuum**
   Robot is exposed to high vacuum such as in space.
6. **Explosives**
   Robot identifies, aims, and destroys explosives such as bombs hidden in enclosures (for example, suitcase).

**Director, Robotics and Automation Laboratory, University of Toronto, and President, Engineering Services, Toronto, Canada.*

7. **Land mines**
   Robot scans, locates, and warns about underground conventional and improvised plastic explosive land mines.
8. **Frequency jamming/security of communication**
   A robot on special operation to attend to an emergency task should be free from any interference in its functioning, command, and control.
9. **Exposed power cables**
   Robots operating on tasks involved in the vicinity of exposed power cables, for example, to prune branches of trees around exposed power lines.

CRITICALITY OF SYSTEM DESIGN

The critical design issues involved and requirements of the robotic system to overcome each of the above hazardous environmental conditions are:

- Sealed/waterproof joints,
- Combination of wheels, tracks, and special gear box transmission for robust navigation,
- Radiation hardening of robot components,
- Research on navigation of totally enclosed box type mobile robot by internal Inertial Propulsion,
- Fire proof enclosures,
- Components which are robust to degassing,
- Laser guided explosive detection, aiming, diffusion of bombs by water jet disrupters,
- Adaptive control of robot to maintain safe distance of mine detector from the Improvised Plastic Explosives, detection and diffusion technology,
- Wireless communication between robot and base station which is robust to terrorist interference and deciphering, and
- Robot flexibility to operate around exposed power cables; robustness of robot not to get damaged if it touches a cable.

APPLICATIONS

After conducting extensive research in the field, robotic systems have been developed for specialised environments and tasks. Examples are:

- Re-configurable robotic arm for space applications (for use by Canadian Space Agency).
- Mobile robot for laser guided explosive aiming and destruction of explosives (for use by Royal Canadian Mounted Police).
- Mobile robot for detection and location of underground explosives (for Defence Research Establishment, Suffield, Canada), Fig. 9.1.
- Hazardous and nuclear material handling robot with extended arm for long reach (for use by Atomic Energy Commission, Canada), Fig. 9.2.
- Totally enclosed box type mobile vehicles without any wheels or legs driven by Inertial Propulsion (for Energenius, Canada), Fig. 9.3.
- Forestry manipulator for pruning branches of trees around high tension exposed electric cables (for use by Ontario Hydro, Canada).
- A mobile robot on a hilly terrain, Fig. 9.4.
- A mobile robot climbing up the steps, Fig. 9.5.

Fig. 9.1. Mobile robot for explosive detection and disposal

Fig. 9.2. Modular extension arm for long reach

Fig. 9.3. Remotely operated loading into van

Fig. 9.4. Application on natural hilly terrain

Fig. 9.5. Mobile robot climbing up the steps

10

Development of Robotic and Virtual Reality Systems for the Betterment of Human Conditions

Dr. M. Vidyasagar, Dr. C.R.J. Prakash Naidu†,*
Dr. P.J. Narayanan‡, Ranjan Kumar Mishra§

INTRODUCTION

A revolutionary change in factory production techniques and management was predicted by the end of the twentieth century. Every operation in the projected factories of future, from conceptualisation to product design, manufacturing, assembly, and product inspection, was expected to be monitored and controlled by sophisticated intelligent robotic systems. However, we see that the projected factories of the future are neither universally implementable, nor always techno-economically viable. The fully automatic factory requires unprecedented involvement of intelligent systems and machines operating in concert in the entire product development process. The technology is achieved, but its application will remain in limited sphere. With the rapid development in the computer technology and programming techniques, the intelligent systems of the future will find more and more applications in fields other than conventional manufacturing factories. Robotic systems are becoming more and more popular and useful in non-industrial and unusual fields of application such as space exploration, undersea exploration, nuclear and hazardous environment applications, and, most importantly medical and human-aid applications. As part of the development in such fields of robotics, IRIS, Bangalore is developing an Intelligent Wheelchair for physically challenged people at a fraction of the cost of current imported systems.

In parallel to robotic systems, Virtual Reality (VR) is also emerging as a major technological breakthrough, which strives to provide users a perception of world that does not actually exist. VR has great applications in design, training and entertainment. A product can be analysed using VR simultaneously for its functionality, compatibility with other parts, ergonomics, aesthetics, etc., as soon as its preliminary design is done. This not only reduces the costs due to the shorter design cycle time but also helps in satisfying the multiple constraints right from the start. VR has become an important training tool not only for pilots and tank commanders, but also in industrial settings and even in medicine. For example, no systematic method exists

* *Director, Center for Artificial Intelligence and Robotics (CAIR), Bangalore, India.*
† *Head, Robotics and Automation, Center for Artificial Intelligence and Robotics (CAIR), Bangalore, India.*
‡ *Head, Vision and Virtual Reality, Center for Artificial Intelligence and Robotics (CAIR), Bangalore, India.*
§ *Scientist, Institute for Robotics and Intelligent Systems (IRIS), Bangalore, India.*

today to train surgeons on minimally invasive or endoscopic surgery. Virtual reality can provide the trainee surgeons with not only the visual cues as would be seen through the keyhole camera but also the force feedback as would be felt through the surgical instruments. CAIR and IRIS are currently developing a laparoscopic surgery training system that integrates models of human abdomen, force characteristics of abdominal tissues, force-feedback devices, etc., to train surgeons in laparoscopic procedures as well as to evaluate their performance in it. CAIR and IRIS also plan to develop better laparoscopic surgery instruments using the knowledge gained from this exercise. VR, with its integration of multisensory stimuli, can also play an important role in helping children with learning disabilities to overcome them. VR has considerable scope in entertainment and education. It is, for instance, possible to teach the fundamental concepts of physics through entertainment using VR by making the virtual world behave contrary to the real one.

OBJECTIVES

The aim of the developments discussed is deployment of robotics and VR systems for the betterment of human conditions. This can be broadly classified into two categories:

- helping people to carry out the normal day to day work schedule; and
- providing better environment for application of skills to carry out specialized tasks.

In the first category, a product like an Intelligent Wheelchair has tremendous potential for people with motor disabilities to carry out their day to day work. Similarly, developing a VR based haptic device will definitely enhance the skills of a Laparoscopic surgeon.

RELATED DEVELOPMENTS AT CAIR AND IRIS

Development of an Intelligent Wheelchair

Development of an indigenous intelligent wheelchair was found desirable to aid physically handicapped people in their everyday activities. As safety and ease of navigation are important issues for the handicapped, the varying levels of configurations are to be developed with different features.

International Status

Efforts are under way in western countries to develop advanced wheelchairs, especially in USA and Europe. In USA, an estimated number of 10,000 motorised wheelchairs are sold annually, each wheelchair costing more than $3,500. Voice activated powered wheelchairs are also under development costing $20,000 depending on the levels of sophistication. In Europe, a group of agencies from six countries are working on a joint project called IMMeDIAte (Integrated system for Mobility and Manipulation for Disabled persons) to develop an Intelligent Wheelchair equipped with a robotic manipulator to aid the physically handicapped.

National Status

It is estimated by ALIMCO (Artificial Limbs Corporation of India Ltd., Kanpur) that 14 lakh wheelchairs are required in the Indian market. This information is obtained from a report available with Mobility India, Bangalore whose source was ALIMCO on the basis of average sales figure ratios. Another statistical data is the number of handicapped in Karnataka which is around 2,00,000 (*Source:* Department of Welfare of Disabled, Government of Karnataka). A subset of these people require wheelchairs.

Our detailed survey reveals that a large number of foreign companies sell motorised wheelchairs with varying levels of sophistication, but they are prohibitively expensive. However, there are no commercial motorised wheelchair producers in India to date.

Importance of the Project in the Context of the Current Status

As there are no motorised wheelchair manufacturers existing in India, the development of this technology and the transfer of this technology to an Indian manufacturer for large-scale production are highly desirable in the Indian context. At present, in view of the high cost of imported motorised wheelchairs (the cost is above Rs. 1.5 Lakh) and the difficulty of their maintenance (no local spares are available), very few people use such wheelchairs.

Development of Prototype Models

The intelligent wheelchair developed at IRIS is a joystick-controlled motorized wheelchair with obstacle detection system. During the process of development, IRIS has designed and tested various drive systems: (1) Direct-drive, (2) Power-base, and (3) Friction drive.

The friction drive option is retrofitted on an existing manual wheelchair and a pair of motor-gear sets drives the large wheels of the wheelchair through non-metallic friction wheels (Fig. 10.1). The Direct-drive option uses smaller wheels that are directly driven at the axles. The power-base option is variant of the modular direct-drive option in which any desired chair can be fitted over the base platform (Fig. 10.2).

All the three types of drive options offer features, namely joystick based controller, fast and slow speed modes, safety sensors for automatic detection of obstacles and subsequent stoppage of the wheelchair, visual and audio indication of battery low state and on-board battery charger. The embedded joystick controller is flexible and adaptable to incorporate user specific functional tuning of parameters, such as sensitivity and response of joystick, speed limits, etc.

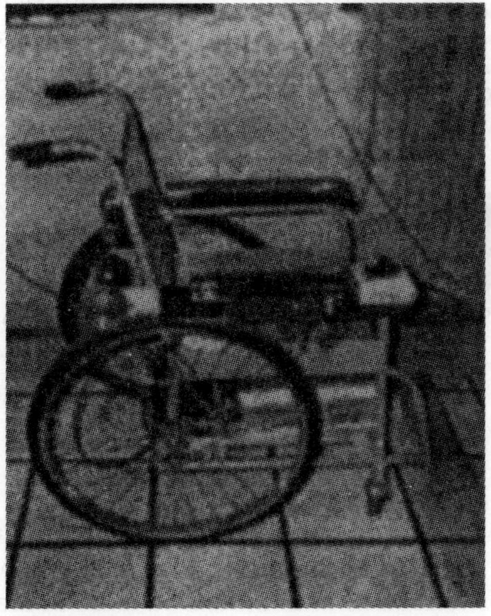

Fig. 10.1. Friction drive wheelchair

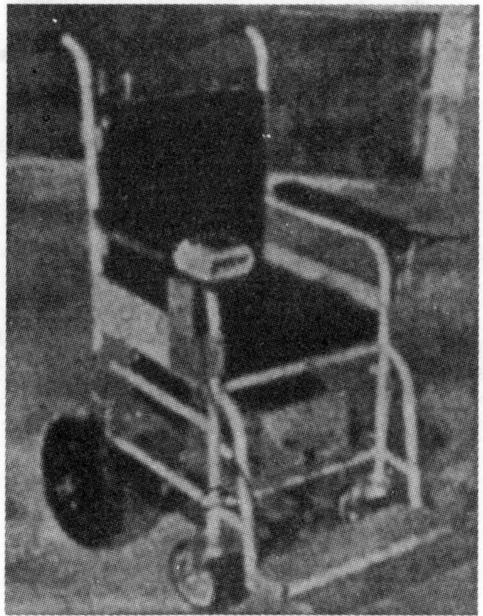

Fig. 10.2. Modular direct-drive wheelchair

All these features are provided at a very economical target price of about Rs. 35,000 to Rs. 45,000 as against Rs. 1.5 lakhs to Rs. 2 lakhs basic price of an imported wheelchair. Use of the indigenously manufactured DC motors and in-house development of compatible controller and drive electronics make the price of the wheelchair affordable. The wheelchair is currently under field trials. The wheelchair can be mass-produced as a commercial product for the market with the required ergonomic and aesthetics improvements, which are under process. After the field trials are completed and improvements are implemented, the cost advantage will make the wheelchair globally competitive and ready to penetrate the international market. Additional features currently being researched include integration of ultrasonic sensors for automatic obstacle detection, speed reduction and stoppage, and speech recognition based activation of wheelchair. Integration of a dexterous robotic arm to help specific users to conduct day to day manipulation tasks is also under consideration. This will lead the research towards robotic applications in domestic and household works.

Development of VR-based Laparoscopic Training System (Fig. 10.3)

Laparoscopy is a minimally invasive surgical technique developed in mid-eighties, which has greatly reduced the time necessary for the patient to heal after surgery. As an example of the technique, consider a minimally invasive gallbladder removal procedure, called laparoscopic cholecystectomy. Instead of making a large incision across the abdomen to expose and remove the gallbladder, the surgeon makes three narrow incisions in a row near the navel. Tools to perform the operation are inserted through trocars into the holes on either side and a camera attached to a lens system in a tube is inserted through the middle opening. This allows the doctor to separate the gallbladder from the liver, connecting ducts and blood vessels, and

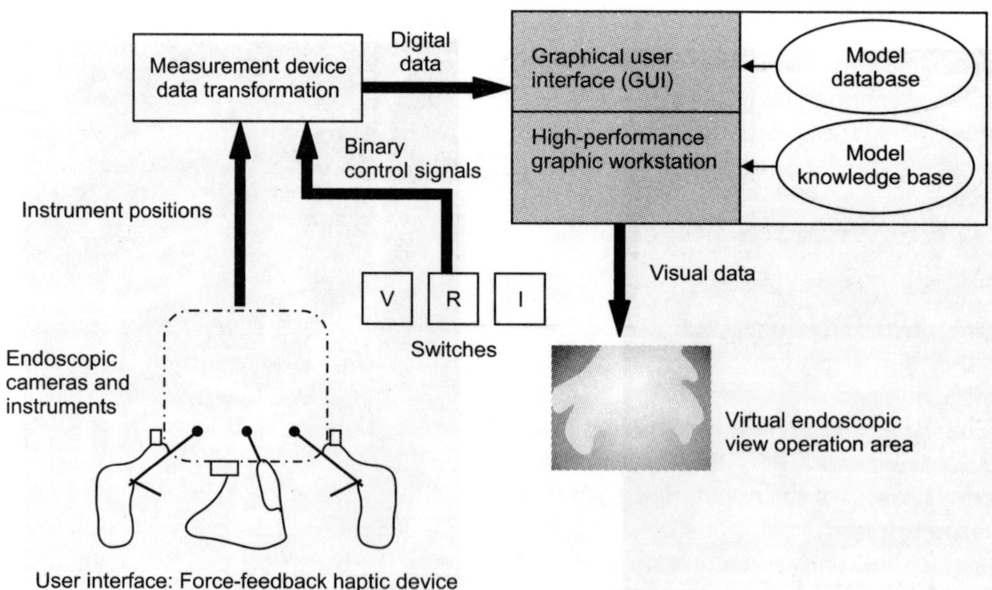

Fig. 10.3. Block diagram of the setup for the VR-based laparoscopic training system

remove it through one of the openings. Because only three narrow incisions are needed to perform the operation rather than a large incision across many muscle groups, the patient needs far less time to recover after the surgery.

However, the new technique presents new challenges for the surgeon. The operating conditions are much more confined than in traditional procedures. The indirect access to the operation area causes a lot of disadvantages for the surgeon: restricted vision, difficult hand-eye coordination, and handling of instruments with limited mobility. Although the abdomen is pumped full of carbon dioxide to create an open cavity in which the surgeon can work, it requires some practice to recognize the anatomy from a new viewpoint provided by the camera. Also working with the laparoscopic surgical instruments can be more difficult than using more traditional tools, because the laparoscopic devices are inserted into holes that allow the tools to pivot and be pushed in or pulled out, but do not allow the tools to be placed at arbitrary angles and locations during the surgery. All these factors combine to make laparoscopic procedures good candidates for computer-based simulation.

Concept of the VR-Based Training System

The basic structure is called a 'man-in-loop' simulation because the human who acts as user or trainee closes the control loop. The system is divided in two parts: the real world that represents the user's immediate environment and the virtual world created by the simulation with hardware, software and data. The virtual world is again divided into two main systems: Force-feedback Haptic System and Virtual Simulation System.

CAIR and IRIS are developing a VR-based force feedback haptic device. The generalised haptic interface device under development at CAIR and IRIS is controlled by VR simulation software.

The user manipulates the instrument; the simulation software tracks the position and orientation and generates real-time graphics and realistic feel. As the instrument interacts with virtual organs, tissue, or objects, the VR simulation commands force of the haptic device to create the real feeling. The device under development will simulate surgical tool(s) having five degrees of freedom with three degrees of freedom force-feedback, just like a laparoscopic tool inserted through a trocar into the abdomen. The tool can be pivoted around an area roughly the shape of a half-sphere. The tool can also be moved further into or out of its virtual hole, or the tool can be twisted around its longitudinal axis. The fifth degree of freedom will allow the user to open or close the virtual tool like a clipper.

Force Feedback Haptic Device

The three degrees of freedom haptic device, being developed by CAIR and IRIS, accommodates free arm and hand motions about the shoulder as the centre of rotation (Fig. 10.4). The transmission is based on cable-capstan drive trains. All the three motors are having position sensors. The base motor primarily responsible for the left-right motions, or X motions is grounded. A cable transmission converts the motor rotations to the rotation of a large disk, which carries the rest of the mechanism. The second and third motors responsible for up-down and in-out motion, or Y and Z motions, ride on a ring mounted on the large disk, using similar cable-capstan drive. The second and third motors control the position of two linkages in a four-bar parallel linkage design. The sensor motor packages measure the endpoint position and provide force feedback in three translational degrees of freedom.

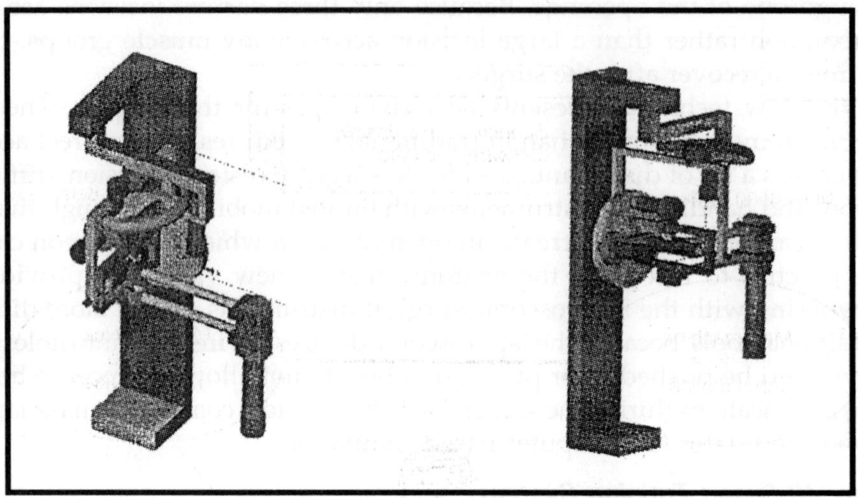

Fig. 10.4. CAD model of the Force-feedback Haptic Device under development at CAIR and IRIS

A three axes servo controller drives the device. The software and the controller maintain a high servo rate ensuring stable close-loop control of the device. The software handles the device kinematics and other robotic calculations like high level position and force calculations. The software generates the haptic effects based on geometry (point haptic exploration), or force-time profiles (sinusoidal vibrations and jolts), or user defined force fields based on the instructions from the simulation software.

Virtual Reality System

The Virtual Reality system integrates computer-generated models of human abdomen, force characteristics of abdominal tissues, and the force-feedback devices. This enables a doctor's surgical interaction with the computer-generated simulation of a patient's body. Doctors or scientists can navigate through these virtual models for mock surgery training. Virtual surgery software allows a trainee surgeon to practice a procedure on a computerized simulation until competence is achieved. Virtual reality software relies on fast graphic processing, imaginative input devices, and manoeuvrable robotics or haptic devices. The simulation software also requires accurate geometrical models of anatomical structures along with realistic dynamic physical models for various tissues so that the simulation not only looks realistic but also behaves realistically (Fig. 10.5).

For modelling and realistic simulation, following data is required:

- Model database
  Model database defines the geometrical shapes and the physical/mechanical properties of the tissues, organs and vessels as well as the geometry and kinematics of the instruments.
- Knowledge database
  Knowledge database specifies the interaction behaviour and handling of the model manipulation. This can be divided into three steps:

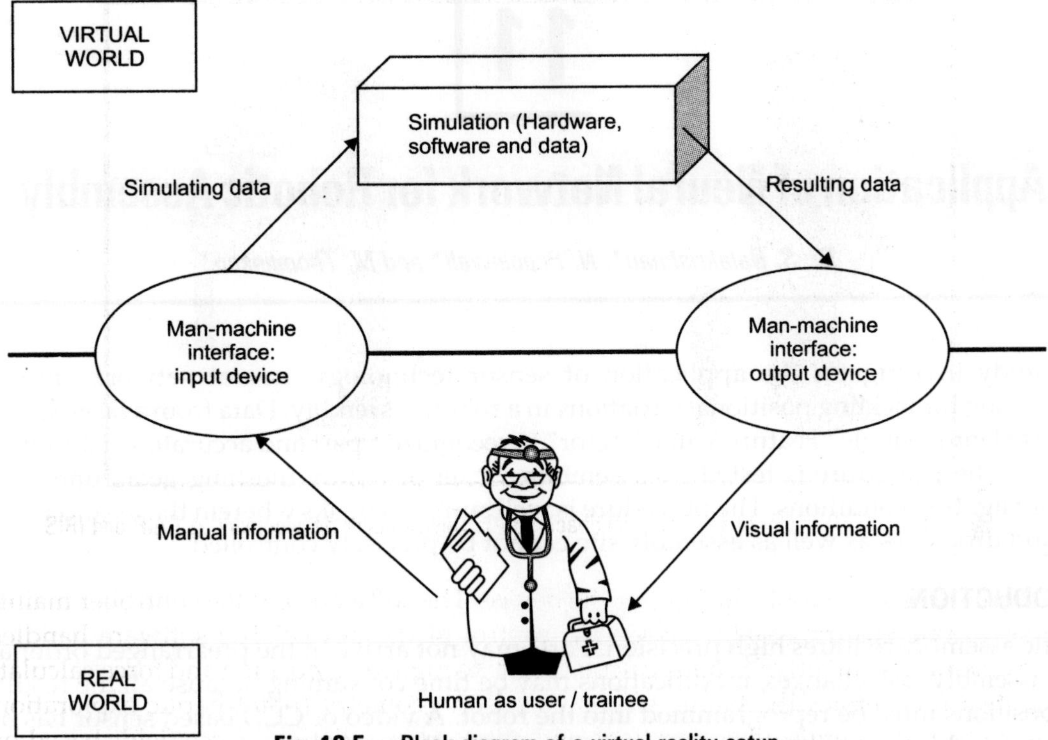

Fig. 10.5. Block diagram of a virtual reality setup

1. Collision recognition: checks for space sections simultaneously used by two objects
2. Interaction management: handling of different interaction types and evaluation (depending on type of end-effector)
3. Model modification: basic manipulation of the deformable object, change of the model structure and efficient data structure.

The outcome of the research on mobile robotics at CAIR and IRIS resulted in the design of a low cost intelligent wheelchair that is a globally competitive product. The technology is being transferred to the potential manufacturer(s) for mass production.

The experience gained and technology evolved from the development of VR-based systems at CAIR and IRIS is resulting in the design of a laparoscopic surgery training cell.

The application of robotics and VR in the future will broaden in the areas outside the realm of conventional manufacturing shops.

11

Application of Neural Network for Robotic Assembly

Dr. S. Balakrishnan, N. Popplewell* and M. Thomlinson**

This study investigates the application of sensor technology, neural network and back propagation for tackling positional variations in a robotic assembly. Data from a laser sensor is converted into a single "Feature Value Vector" to recognize a part and accurately determine its location. The procedure is tested by assembling a set of tightly meshing gears under poor ambient lighting conditions. The procedure is well suited for tasks wherein the locations of the pick and drop sites as well as assembly site cannot be precisely controlled.

INTRODUCTION

Robotic assembly requires high precision. Parts may not arrive in the prearranged order and, if the assembly task changes, modifications may be time consuming because all the resulting new positions must be reprogrammed into the robot. A video or CCD based sensor has been used mostly to generate a two-dimensional image of an object. Data obtained from ultrasonic sensors in combination with Artificial Neural Network (ANN) have been utilized to identify the letters of the English alphabet with 99 per cent success. An ANN's ability to determine distances and orientations, which is essential in assembly, has not been investigated previously.

EXPERIMENTAL SYSTEM

An inexpensive but reliable laser range sensor [Model # LAS-8010V from Adsens Technology, Canada], is adopted here because it poses fewer computational demands and is unaffected by lighting conditions. The sensor is attached to the end effector of a modified CRS M1A manipulator. It measures the distance between a laser source and an object interrupting the emitted beam so that a one-dimensional "image" is formed. A three-dimensional pattern is constructed by transporting the sensor around the object. The D-H model of the manipulator was programmed into an IBM compatible microcomputer. A custom-developed six axis motion control interface links the PC and the robot. A schematic of the laser range

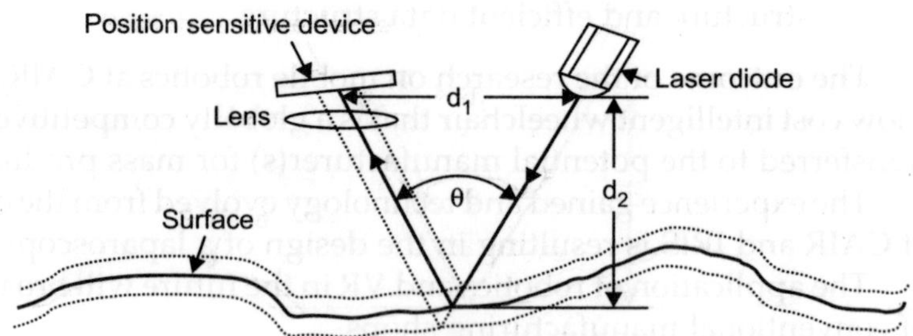

Fig. 11.1. Triangulation laser

**Department of Mechanical and Industrial Engineering, University of Manitoba, Winnipeg, Canada.*

finder's operation is shown in Fig. 11.1. Here d_1 is the known lateral distance between the laser's outgoing light and that returning to the position sensitive device. If θ is the previously fixed angle between these two rays, then the normal distance to the object, d_2, can be found using

$$d_2 = (d_1 \cot(\theta/2))/2 \qquad ...(1)$$

A voltage in the range 1 to 5 V is produced by the sensor. An analog to digital board digitizes this voltage. Details of the control strategy can be worked out. By applying the standard D-H forward transformation, the point of contact of the laser light with a reflecting object's surface, (P_{6x}, P_{6y}, P_{6z}), can be found.

Multi-tasking and coordination of all control programmes was achieved using QNX™. QNX is a multi-tasking operating system, like UNIX, that is capable of pre-emptive scheduling. The data acquisition rate, the scanning pattern and speed, the microcomputer's CPU speed and capacity, the size and orientation of the features as well as the influence of measurement noise were considered in the design and proper selection of the various control parameters.

RECOGNITION

A procedure has been developed to recognize a part by employing an ANN and back propagation learning to identify rotation invariant and feature values. During the learning process, an input state vector is presented and propagated forward to determine the output signal. The output and target vectors are compared which results in a difference or error signal that is propagated back through the network in order to adjust the coupling strengths or weights. This propagation of error effectively teaches the network, the relationship between the input and output vectors. The goal of the back propagation algorithm is to minimize the error, E, through a least-mean-square (LMS) procedure.

This method has been implemented in software to process the laser data retrieved during a scan across an object. It provides a means to teach the network to identify a pattern of range values corresponding to a particular object. A feature value vector (FVV) is calculated by using the binary output from the laser sensor that correspond to a series of constant width, annular sections which wholly or partially intersect the part's superimposed image. A typical superposition is illustrated in Fig. 11.2. To be rotation invariant, the origin, 0, of the polar

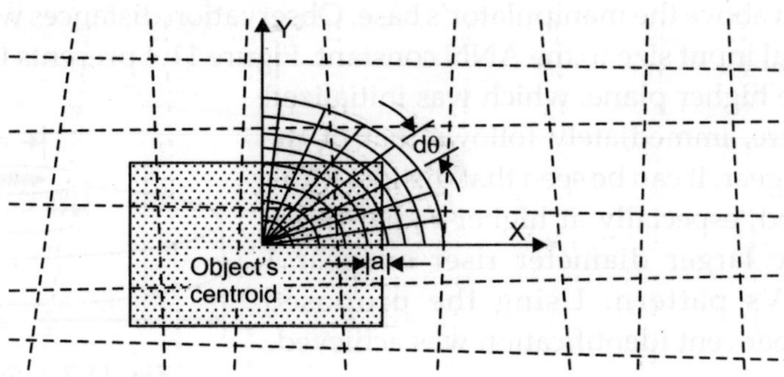

Fig. 11.2. Plan of scan path

co-ordinates describing the annuli should coincide with the part's centroid. If not, the feature value is recalculated by using the initially computed centroid as the origin. A FVV, $s(r, z)$, is represented by

$$s(r, z) = \int_{D(r)} f(x, y, z)dx\, dy \qquad ...(2)$$

and $$D(r) = \{(x, y); r^2 \leq x^2 + y^2 < (r + a)^2\} \qquad ...(3)$$

Here $f(x, y, z)$ describes the part's three dimensional image and $D(r)$ is a typical annulus having an empirically found width, a = 1 mm. If material is found in a given sector above a specified horizontal plane $f(x, y, z)$ becomes one. If no material is found, $f(x, y, z)$ is set to zero for that sector. Then the integration of equation (2) reduces essentially to a radially weighted sum of the unity values. This summation is stored and the procedure is repeated for annuli starting at different radii, r.

Different parts have locations where the amount of their material differ. Consequently the set of FVVs for a part should have distinctive characteristics even though the part may not be represented precisely. For example, the radially weighted summation will obviously exaggerate the influence of material at a greater radius. Furthermore, a denser scan will intuitively generate a correspondingly larger FVV because of the proportionate increase in the number of data points describing the image. Hence, the scan density should be invariant. A moving five point average is used to smooth variations whilst preserving the overall form of a FVV.

FVVs for a part constitute the input to an ANN whose output is a strength of recognition. Calculating the total sum of the squared nodal differences (TSS) for the network compares input and output. Then the nodal connections are modified by using a non-linear sigmoid and the process is repeated until the TSS is below that specified by the user.

The ability to recognize parts was assessed initially by employing five different gears that were readily available. Three ANNs were utilized which always had 88 inputs, 5 outputs but 10, 15, or 20 hidden nodes. Sets of FVVs were calculated for five different combinations of location and orientation per gear over a constant horizontal radius that spanned the largest gear. The resulting 25 sets of FVVs were used to train the ANNs and gauge their performance. The key dimensions of the five steel gears are presented in Table 11.1 with respect to the nomenclature given in Figure 11.3. The FVVs were recomputed in two horizontal planes at 2 mm as well as 9 mm above the manipulator's base. Observation distances were halved in each plane to keep the total input size to the ANN constant. Figure 11.4 presents the resulting FVVs where the set for the higher plane, which was initialized again at a gear centre, immediately follows that of the lower plane for each gear. It can be seen that distinct FVVs forms were produced, especially at higher radial sector numbers where the larger diameter riser on gear 1 prolonged the FVVs pattern. Using the proposed methodology a 100 per cent identification was achieved.

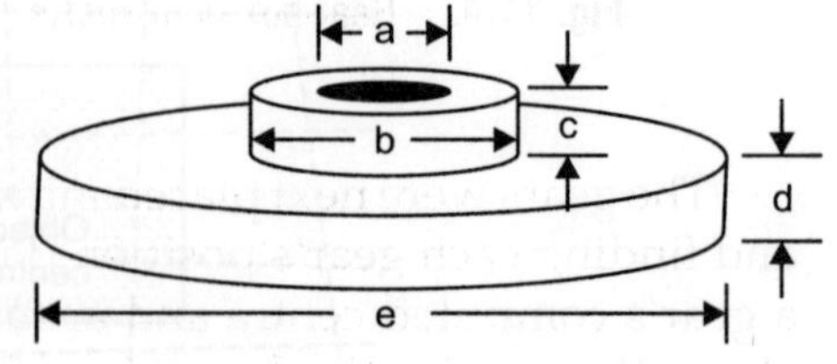

Fig. 11.3. Gear parameters

Table 11.1. Gear dimensions

Gear No.	a (mm) ϕ	b (mm) ϕ	c (mm)	d (mm)	e (mm) ϕ
1	10	34	15	4	70
2	10	20	8	5	72
3	10	15	4	6	49
4	10	15	4	6	39
5	10	13	2	8	28

ASSEMBLY

Assembly involved recognizing and grasping the gears before placing them on five correspondingly numbered, cylindrical shafts chamfered at their ends. The shafts' arrangement is shown in Fig. 11.4 whilst Table 11.2 relates the centres of the remaining shafts with respect to that of shaft one. The casing supporting the shaft had to be within ±5 mm of a preselected position, regardless of orientation, for at least two shafts to be within the manipulator's limited workspace. Correctly locating two shafts (and their centres) enabled not only the remaining shafts' centroids but also the position and orientation of the rectangular casing to be defined because shaft separations were distinct. A check was provided by tightly scanning two 20 mm square areas just above the tallest riser to confirm the calculated locations of shafts 1 and 5.

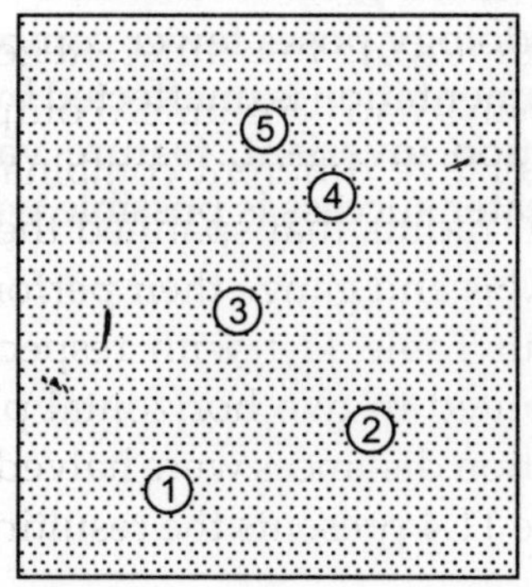

Fig. 11.4. Gear box with shafts

Table 11.2. Shaft offsets with respect to shaft 1

Gear No.	$r_{1,A}$ (mm)	$\theta_{1,A}$ (rad)
2	67.4	1.984
3	80.2	2.778
4	122.2	2.761
5	141.5	2.957

The gears were next placed individually within the scan field before searching, recognizing and finding each gear's position (Fig. 11.5) The manipulator's end effector was moved above a gear's computed centre and around its riser for better grasping. The gear was transported above the previously determined location of the correspondingly numbered shaft, lowered

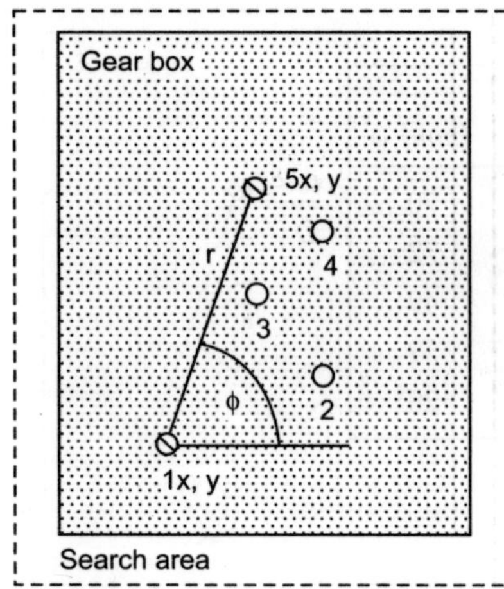

Fig. 11.5. Gear box in search area

Table 11.3. Assembly success rates

Gear No.	*Correct Assembly*	*% of assemblies, correctly executed*
1	115	95.8
2	83	69.2
3	119	99.2
4	120	100
5	119	99.2

gently and released. Then the manipulator returned to its home position and the process was repeated. However, if an adjacent gear had been assembled previously, the next gear was lowered only to where its bottom face was parallel to the top of the adjacent gear. Then it was twisted gently back and forth to facilitate meshing as the gear was lowered further. Table 11.3 gives the results of 120 tests in which the gears arrived in arbitrarily selected sequences. The average success was just over 92 per cent primarily because of slight shaft misalignments sometimes preventing extremely small gear clearances.

It is demonstrated that recognizing, and locating preselected parts is feasible by using an inexpensive laser range finder. The neural network methodology poses fewer computational demands than standard, vision based sensing and the instrumentation is unaffected by lighting conditions. The neural network normally identified rotation invariant, feature values and assemble gears arriving in widely different sequences. Greater success is possible with a well functioning manipulator.

12

Technology of Water Jet in Manufacturing and Machining

*Dr. B.V.A. Rao**

INTRODUCTION

The fluid jet is a versatile tool for many industrial applications. The most common fluid used for making a powerful jet is water. Water jet is a thin stream of water ejected from a tiny orifice, connected through high pressure hoses from a water pump. Depending on the pressure of operation, the speed of the jet can reach up to 5 times the speed of a commercial jumbo jet. Water jet can virtually cut through any material without any friction, wear, heat and dust.

If abrasive particles are added to the stream, it is called an abrasive-entrained water jet or Abrasive Water Jet (AWJ). There are many ways of adding abrasive particles to a fluid jet. The AWJ has now become a standard tool for many machining applications (drilling, milling, turning, thread cutting in sensitive materials such as glass, composites, etc.).

Cavitating Jet

If water vapour bubbles are induced in the jet, then it is called a cavitating jet. This type of jet is very useful for cleaning and fragmentation applications. This type of jet operates at fairly low pressures and is becoming increasingly popular for many operations under submerged environment (for example, *in situ* cleaning of oil wells).

Pulsed Jet

If the jet is broken into slugs of small lengths, then it is called a pulsed jet. Operating at fairly low pressures, this type of jet is very powerful and can be used to enhance the cutting, cleaning and fragmentation applications. Mining of hard rocks is an example.

The applications ranging from ordinary cleaning to precision manufacturing (for example, the panels of an aeroplane) and, more recently to highly sophisticated medical surgery (liver, gall bladder, brain tumour, etc.). It has a great potential for recycling of waste materials, such as tyres and, for agriculture (highly efficient method for applying herbicides, resulting in substantial savings in costs). Used properly, the technology is environment friendly and user friendly.

The main advantage of the water jet process (with or without abrasives) are as follows:

- Virtually any material can be machined.

*Pro-Chancellor (R&D), Vellore Institute of Technology, Deemed University, Vellore, Tamil Nadu, India.

- The process is essentially 'cold' so that there are no thermal influences thus avoiding changes in the material properties.
- Jet reaction forces are relatively low allowing easy and accurate manipulation by robots or multi-axis tables.
- The jet is a precise tool which can cut in all directions (multidirectional), and normally does not deform the material during cutting.
- The jet can be of a suitable diameter to give small kerf widths when cutting.
- Negligible production of dust.

Water jet machining is no more a myth and a curiosity, and there are 24-hour production lines which have established hardware reliability, thus convincing that this tool is no longer 'non-conventional'; optimization of the process depends upon the hardware demands created by proper orifice sizing, pressure requirements, mechanical fatigue, etc. It is necessary that all the key personnel should have a thorough understanding of the theory and mechanics of the systems.

WJM is similar to LBM, and EBM.

A given amount of energy is concentrated on to a very small point in the work material to cause material removal.

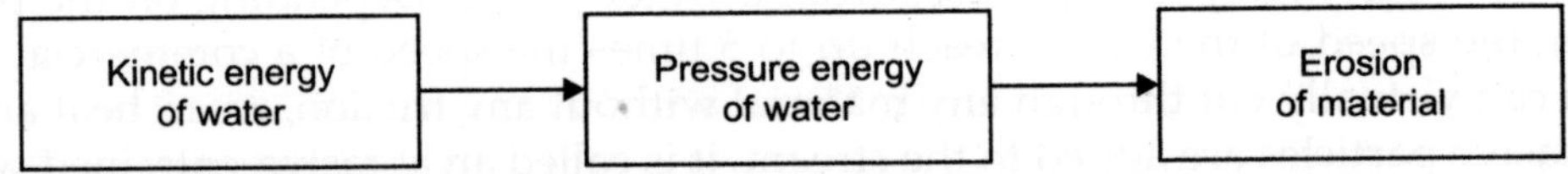

The erosion of material is caused by

- Localized COMPRESSIVE failure of the work material,
- SHEAR failure caused by high speed radial flow,
- SPALLING, and
- CAVITATION.

WJM employs a

high pressure (1500–4000 MN/cm^2), and
high velocity (up to twice the speed of sound).

Jet of water, which, when bombarded on a target, erodes the material. Fine machining of work materials are possible.

High Kinetic Energy leads to very high water pressure on the work material. When water pressure exceeds strength of the material, erosion takes place.

JET THEORY (Fig. 12.1)

Nozzle converts high pressure water to high velocity water jet.

In Eqn. (1), V_a ⇒ Small
P_b ⇒ Small
Friction loss ⇒ Small

Therefore, $V_b = \sqrt{2P_a / p}$

$V_{\text{actual}} = \text{Cd}\ V_{\text{theoretical}}$

Cd = Coefficient of discharge.

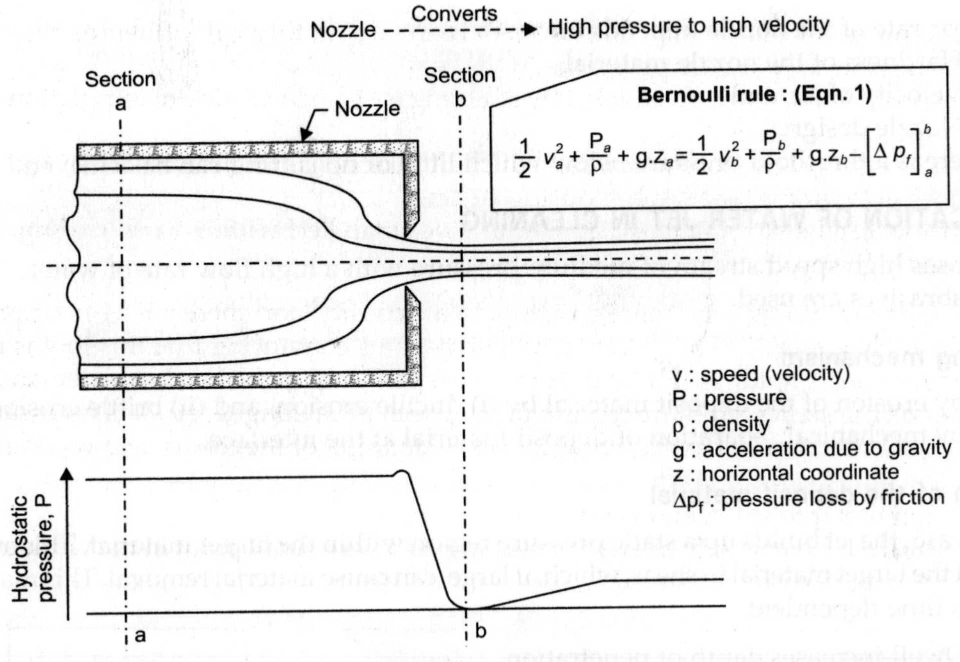

Fig. 12.1. Jet theory

Interaction of jets with their surroundings

1. Coherent jets will be destroyed partially.
2. Divergence of jets occurs. This reduces specific energy input.
3. Loading changes from static to dynamic, and
4. Energy is absorbed by surroundings.

Influencing Factor	Interaction of jet with the material
• Medium of jet	- removes surface layer
• Geometry of the nozzle	- influences substratum layer
• Entering medium	- removes substratum material.

PROCESS CRITERIA AND PARAMETERS

Cutting medium in WJM is water and abrasives.

Material Removal Rate (MRR) depends on

- Mass flow rate - Nozzle diameter.
  - Fluid pressure
- Velocity - Fluid pressure
- Stand Off Distance (SOD).

Geometry and finish of the workpiece depends on

- Nozzle design
- Jet velocity
- Cutting speed (i.e. traversing speed) and depth of cut, and
- Properties (especially hardness) of the material being cut.

Wear rate of the nozzle depends on

- Hardness of the nozzle material,
- Velocity of jet, and
- Nozzle design.

There is a threshold pressure below which little or no cutting can be achieved.

APPLICATION OF WATER JET IN CLEANING

- uses high speed stream of medium pressure with a high flow rate of water,
- abrasives are used.

Cleaning mechanism

- by erosion of the deposit material by (i) ductile erosion, and (ii) brittle erosion
- by mechanical separation of deposit material at the interface.

Erosion of the deposit material

In this case, the jet builds up a static pressure region within the target material. This stress puts areas of the target material in shear, which, if large, can cause material removal. This macroscopic event is time dependent.

- Dwell-increases depth of penetration,
- Movement of jet-increases eroded area,
- Loading of the jet (can be adapted to material properties)
  1. Static with a coherent jet, and
  2. Dynamic with a jet consisting of droplets.

Mechanical separation of the deposit material from substratum

This depends on initial erosion of the depth of the deposit to expose the interface.

The mass removal takes place by the buildup of hydrostatic pressure.

Rate of cleaning depends on manipulation of the jet and the effective area of the jet on the target metal, whether it is

- Rotary or oscillatory manipulator, or
- Spreading jet caused by a machined slot at the outlet nozzle depending upon either fan shaped jet or cone shaped jet.

Cleaning with abrasive water jets

- enhanced performance
- much lower pressures are used
- abrasive jets are used for cutting and cleaning.

Mechanism of erosion depends on type of material leading to brittle fracture or ductile fracture.

APPLICATION OF WATER JET IN CUTTING

Parameters involved in abrasive water jet cutting are:

- Water pressure and flow rate,
- Stand off distance (SOD),

- Cutting speed,
- Nozzle diameter,
- Length of nozzle, and
- Type of grains, grain size and abrasive flow.

Other machining operations

- Water jet machining is the most recent non-traditional machining technique,
- WJM can be used for machining of metals and non-metals,
- Compatible and economical with materials like ceramics, ceramic composites, metal matrix composites, laminates and fibre reinforced resin composites, and
- Used for cutting rocks and concrete.

Turning

- Material removed in the form of very fine debris
- Abrasive size determines surface finish left on the job.

Milling

- Depth control is difficult.

Drilling

Hole shape obtained in drilling is affected by

- Jet structure
- Target material
- Stand off distance (SOD).

Engraving and marking

- Done with precision jet at high traverse rates or at relatively low pressure.

Construction work

- Used for pre-treatment of stone, concrete and metallic surfaces
- Used to remove paint, bitumen, coating material, rust, etc.

COMPONENTS OF WATER JET MACHINING SYSTEM

(1) Pumps

- High pressure pumps are used
- Pressure intensifiers—combined together to get sufficient flow (pressure up to 4000 bars achieved)
- Diaphragm pumps—used for lower pressure operation like drain cleaning (1000 bar maximum pressure).

(2) Nozzles

- Made of hardened steel or tungsten carbide for low pressure applications
- Sapphires, diamonds for high pressure applications held in a metal nozzle holder; (Diameter of nozzle = 0.08–0.60 mm).

(3) Hoses, tubes and fittings

- Flexible hydraulic hoses are used for low pressure applications up to 2000 bars
- Rigid tubes for high pressure applications (burst pressure should be at least 2.5 times working pressure)
- Lining is made of synthetic rubber
- Reinforcement to plies of spirally wound high tensile wire or synthetic yarn
- Outer cover is plastic or rubber.

(4) Arrangement of catchers of residual water

- 75 per cent energy is retained by the stream after cutting
- Catchers
  - Room catchers—enclosure of the facility
  - Stainless steel tank catchers—filled with water and steel balls or gravel
  - Compact catchers—small containers with energy absorbing media
- Used for robotic applications
- Dissipates the remaining energy.

(5) Arrangement for waste slurry handling

- Collects and disposes of the spent abrasives, debris, and kerf materials, i.e. materials removed from the workpiece. Waste slurry is transferred to a waste separation and disposal bin.

(6) Gantry arrangement

Water jet machining equipment are provided with gantry arrangement with precision ball screws so that water jet nozzle can be precisely moved and located at any place above the workpiece.

POTENTIAL APPLICATIONS

Generally used in cleaning, cutting and 3-D machining.

CLEANING APPLICATIONS

(1) Production shops:	-	Deburring and cleaning of castings (not very economical).
(2) On-site, field applications:	-	Cleaning of grids and concrete surfaces to remove layers of materials deposited.
(3) Internal cleaning of heat exchangers of chemical plants	-	Cleaning of pipes done with the aid of 3 or 5 rigid water jet assemblies with nozzles.
Internal cleaning of tanks in chemical plants	-	Holding tanks, production tanks, autoclaves, reactors, agitator tanks, containers, etc.
Operating pressure	-	Up to 1000 bars
Flow rate	-	50 litres/min.
(4) Cleaning of pipes and sewers	-	Pipe cleaning head is provided with several nozzle inserts
Pressure	-	Up to 1500 bars
Flow rate	-	90 litres/min.

Paint removal from aircrafts panels

- aquastripping of paint is ecologically safer and faster process.

Hull cleaning

- cleaning of sides and bottom of ship hulls.

Hydrodemolition

- demolition of canals for reconstruction and relaying.

CUTTING APPLICATIONS

Plain water jet

- Food industry: cutting cakes, fish sticks, etc.
- Cutting and sectioning of printed circuit boards
- Advantages
  - Dust free
  - No surface pressure, shocks, bending load, microcraking
  - Suitable for automation.

Abrasive water jet

- Machining of nozzle blade holes in steam turbine and diaphragm spacer rings (replaces EDM technique)
- Precision cutting of 3D-curved surfaces (uses two cutting heads with individual simultaneous control in five axes). Unlike EDM, AWJ has no fire hazard and is environment friendly
- In Aerospace applications: for cutting
  - Composites
  - Honey comb sandwiches
  - Titanium components
  - Nickel and cobalt superalloys
  - Stacked metals
  - Fibre glass reinforced plastics
- Advantages
  - no delamination of fibres from the matrix material
  - no heat damage
  - reduction in tooling costs
  - better fatigue life than sheared parts
- Cutting laminated composites or special glasses (instead of using diamond wire); (process is highly competitive)
- Decorative artefacts made from natural stone, glass, ceramics or metal
- Offshore applications (potentially applicable for deep sea steel pipeline cutting).

MACHINING APPLICATIONS

- Turning
- Spiral thread machining

- Milling
- 2D profile cutting of any shape in thick and tough work materials difficult to cut otherwise.

Special features of water jet cutting

Water jet cutting gives a fine cut with high accuracy, hence most of the intricate jobs can be done by water jet cutting.

Although water jet cutting/machining process is slower than plasma jet and laser jet machining process, it has the advantage of cutting almost all types of materials.

Water jet cutting is mostly suitable for cutting stainless steels of higher thickness as indicated below:

Work material	*Process*	*Maximum thickness of material that can be cut*
Stainless steel	Water jet machining	170 mm
Stainless steel	Plasma arc machining	32 mm
Stainless steel	Laser beam machining	20 mm

Work materials than can be machined by WJM

Although water jet/abrasive jet can cut most of the materials, it is economical and better than other conventional methods, especially in respect of the following materials:

1. Aluminium
2. Ferrous materials like carbon steel and stainless steel
3. Non-ferrous materials like brass, bronze and copper
4. Titanium
5. Special alloys
6. Non-metallic materials like stone, marble, granite, ceramic tiles, glass, rubber, teflon, plastics and leather
7. Composite materials and hybrid materials (Machining of cut-outs and trimming of automotive dash board instrument panels made of composites is a typical application of water jet machining).
8. Cutting of bullet proof glass and tough ceramics.
9. Floor designs in marbles/tiles.

Industry wise applications of WJM

The list is not exhaustive and many new applications are being developed by the respective industries:

1. Aerospace industry
2. Marine/Ship building industry
3. Repair and maintenance work
4. Architectural industry
5. Construction industry
6. Automobile industry
7. Machine tool industry

8. Art and Artefacts (Special designs)
9. Fabrication industry, especially of steel and stainless steel products
10. Interior decoration works
11. Engraving of letters/designs in steel, brass, etc.

CNC water jet cutting

CNC water jet cutting machines have also been developed. The water jet nozzle can be precisely moved along the gantry frame in the X-axis longitudinally using precision ball screws. For this purpose, CNC controller and a suitable CAD/CAM software are used. Software enables the selection of optimum feed for different work materials.

Machine sizes range from 1 m × 1 m to 3 m × 8 m with single or multiple heads of nozzles (water jets). The cutting head provides precise alignment of the water and the focusing nozzle, provides repetition accuracy and high cutting speeds.

Water jet technology is an emerging technology. Generating the required pressures through pumps and nozzles, handling with safety hoses and other ancillaries have given confidence that water jet could be a potential tool in many manufacturing applications. Predictive control enables machining systems incorporating water jets to match or better the accuracy and quality of the competing technologies. Some previously impossible tasks can now be carried out cost effectively opening up the market place for a whole new range of manufacturing materials, and applications. Water jet is environment friendly, and this is closely linked to the continued development of higher efficiency systems. In many cases, application of water jet technology can increase the quality of life for the machine operators by reduction in aerosol emissions, vibrations, noise and dust. This technology can be extended to many different applications.

13

Advances in Thermo-Chemical Heat Treatment Processes

*Keith M. Bennett**

The demands for steel components to perform to higher standards in aggressive environments are ever-increasing and today's engineers are required to produce better quality components—not only in metallurgical properties but also in dimensional standards.

As far as conventional heat treatment processes are concerned, there is a limit to what can be achieved. Omega Technical Services Limited, U.K. has realised this at an early stage in the 1970s itself and has decided to develop their own specialised heat treatment processes to fill the requirements.

The progress made is best summed up by a quotation attributed to Sir Isaac Newton—"One takes the foundation of what has gone before and one builds on it."

It is without doubt that thermo-chemical heat treatment technology has now been pushed into a different dimension. The classification of wear in most situations relates to metal against non-metallic abrasives, metal against metal and liquid or vapour impingement on metals—often with the conditions combining thereby subjecting components to both wear and corrosion simultaneously.

Low temperature thermo-chemical technology can serve to effectively combat these situations.

It is this technology which is being discussed and presented here.

LOW TEMPERATURE THERMO-CHEMICAL TECHNOLOGY

Low temperature thermo-chemical technology can be considered to fall into three basic categories

1. Ferritic Processes
2. Austenitic Processes
3. Ferritic/Austenitic Processes

Many generic processes are available today, particularly those of the ferritic type, but not many in the other two areas. Omega Technical Services Limited, U.K. have developed many technologies pertaining to "Austenitic Processes" and "Ferritic/Austenitic Processes".

**Managing Director, Omega Technical Services, West Yorkshire, U.K.*

Omega Technical Services Portfolio of Processes include:

(a) Ferritic Processes

Alpha	PNC
Alpha	SQ
Seal bond	
Zeta	NT
Zeta	SA
Sealbond	Ultra
Alpha	Ultra

(b) Austenitic Processes

Alpha	Plus 5
Alpha	Plus 10
Beta	25
Beta	40
Beta	PC

(c) Ferritic/Austenitic Processes

Delta	
Delta	SSA
Delta	SSM
Delta	TiG.

FERRITIC PROCESSES

Most commercial ferritic processes are traditionally carried out at a temperature of 570°C but it has been established that a universal temperature of 570°C does not necessarily suit all applications.

The Omega ferritic processes are carried out in the temperature range of 530°C to 610°C giving them greater versatility.

Alpha PNC Process

Alpha PNC, for example, is a pit furnace base gaseous alternative to the original salt bath ferritic nitrocarburising patented proprietary processes, such as tuffiride, tenifer, sulphinuz and sursulf, etc.

The process can be considered to give properties which, at least, match those of the salt bath processes but without the traditional problems (Figs. 13.1 and 13.2).

Salt bath processes involve the use of cyanide salt to varying degrees whilst the gaseous alternative is free of toxicity. The necessity to clean the components by washing or other methods after treatment is completely eliminated and the problems previously experienced with salt residues remaining in blind holes no longer exist.

The immediate surface and sub-surface of all ferritic nitrocarburised components exhibit a certain level of porosity after treatment and this effect is utilised for the retention of lubricants during service. Post heat treatment surface finish suffers to a small extent by the development of the porosity. Gaseous ferritic nitrocarburising results in a comparably lower level of porosity as against salt bath processes, without a deterioration of the lubrication retention capability thereby producing a superior post heat treatment surface finish.

Fig. 13.1. Microstructure of Alpha PNC processed mild steel

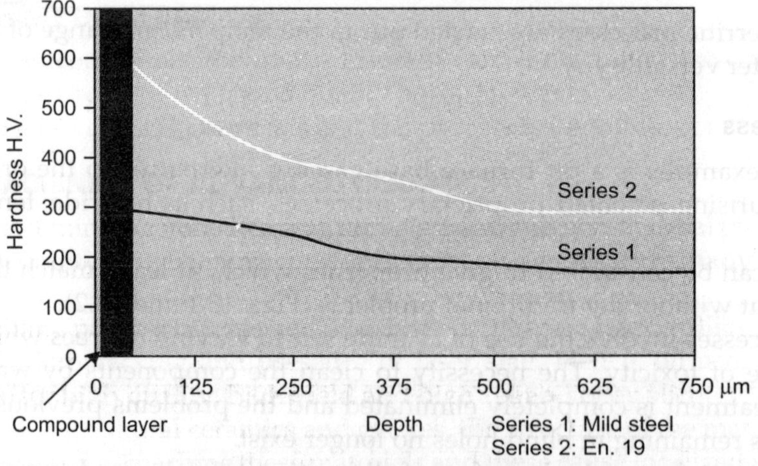

Fig. 13.2. Typical micro-hardness traverses obtained by Alpha PNC3 process

Prior to the development of Alpha PNC process no pit furnace ferritic nitrocarburising process was available and the processing of long components was difficult.

The geometry and size of pit furnaces in which Alpha PNC process is carried out enables the heat treater to process components, such as long crankshafts in a vertical position, thereby reducing the distortion and lowering the operating costs of the equipment resulting in cost savings to be passed on to the customer in most instances.

The response of alloy steels to ferritic nitrocarburising, i.e. the nitriding effect (Fig. 13.3), has been used to great advantage for the processing of fully hardened and tempered hot die steel extrusion and forging dies increasing their service life.

In this connection, the more flexible compound layer at the surface exhibits an enhanced spalling resistance in comparison to the classical nitrided surface.

Ferritic nitrocarburised components can be put into service without any post-cleaning or polishing operations, an obvious advantage over classical gaseous nitriding where the brittle 'white layer' must be removed before components are put into service.

Fig. 13.3. Alpha PNC processing of En. 40B steel

Alpha PNC process, for example, has been demonstrated to meet and improve standard nitriding specifications with savings up to 70 per cent in process cycle times.

The extended turn-round times associated with classical gaseous nitriding are a thing of the past and, as such, the advantage to the customer in today's terms of 'just-in-time' purchasing philosophies, is obvious.

Applications of Alpha PNC Process

- Engine reconditioning (crankshafts, camshafts)
- Gear manufacturing (gears, gearshafts, worms)
- Diesel engine manufacturing (crankshafts, cylinder heads)
- Transmission systems (gears, shafts)
- Pump manufacturing (pump shafts, caps)
- Glass manufacturing industry (mould components, pins)
- Tool and die engineering (press moulds, forming rolls)

Ferritic nitrocarburising can also be carried out in Top Hat and sealed quench furnaces.

Alpha SQ Process

Alpha SQ process is a sealed quench furnace based ferritic nitrocarburising process which is a nitrogen based alternative to the salt bath processes and also to atmosphere processes such as the endothermic gas based nitemper.

Slight growth does occur during the process, as with all ferritic nitrocarburising processes, but it is normally of the order of 0.04 mm on diameter.

The growth is reproducible providing the customer controls the material and machining techniques and, as such, is therefore predictable.

The process is used often for the reclamation of undersize components.

Applications of Alpha SQ Process

- Automobile industry (washers, pins, rollers)
- Aluminium extrusion industry (extrusion dies)
- Tool and die engineering (press tools, forming rolls)
- Engine reconditioning (crankshafts, cam shafts)

It has been demonstrated quite clearly that the compound layers produced at the surface of ferritically nitrocarburised components exhibit excellent corrosion resistance a property which is utilised successfully to eliminate post heat treatment phosphating or plating which reduces manufacturing costs considerably.

Sealbond Process

The Sealbond process developed by Omega Technical Services Limited, U.K. serves to enhance the corrosion resistance further whilst producing an associated aesthetic blue/black finish (Fig. 13.4).

The surface thus produced has a high capacity for lubrication retention during service which is of great advantage for components which operate in positions which are difficult to access for maintenance.

Fig. 13.4. Comparative effect of seawater attack on untreated and sealbond treated surface of H13 welding clamps

Sealbond process competes very successfully with the highly publicised Nitrotec process, e.g. the latest salt spray test results on sealbond indicates a resistance of some 1000 hours against 250 hours claimed by Nitrotec.

Sealbond can also be given a variety of alternative finishes, e.g. enamel paint to enhance the corrosion resistance of a component even further. In this case, the sealbond is being used as a pre-treatment process for painting (Fig. 13.5).

Applications of Sealbond Process

- Chemical Engineering Industries (all moving parts)
- Special Fastener Industries (all fasteners)

Zeta SA Process

A most recent development has been the unique Omega Zeta SA process which was specifically targeted at the surface treatment of previously hardened and tempered low alloy, drop forging dies.

It has always been recognised that for this application, a nitrocarburised surface would be more beneficial than a nitrided surface. Unfortunately, the thorough hardness specification of the forging die is such that the original tempering temperature is much lower than standard nitrocarburising temperatures. If the parts were to be subjected to standard nitrocarburising then the forging die would become over tempered and the thorough hardness would deteriorate.

The Zeta SA process, in fact, is carried out 400°C below standard nitrocarburising temperatures.

Applications of Zeta SA Process

It should be noted that the forging dies greatly benefit from this surface treatment, increasing the length of the production run on an average, by almost 300 per cent. In addition, the

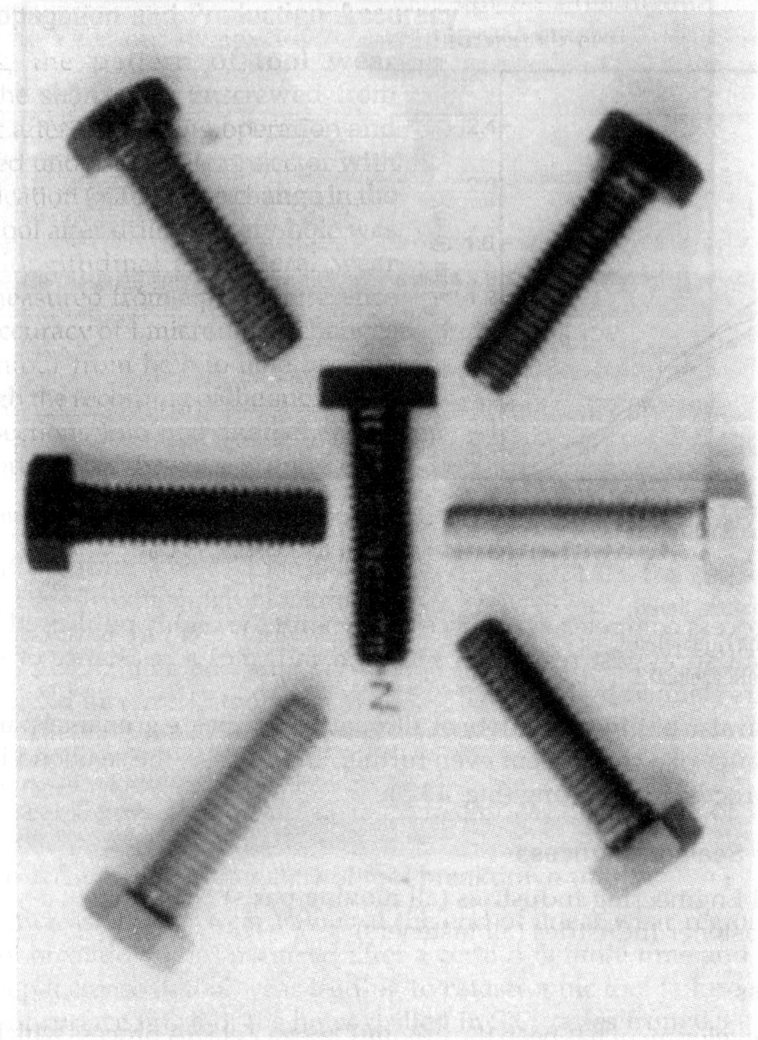

Fig. 13.5. Sealbond/enamel processed fasteners

compound layer produced by the process eliminates pick-up problems enhancing the surface finish of the forged product (Fig. 13.6).

A NEW GENERATION OF THERMO-CHEMICAL HEAT TREATMENT PROCESSES

The majority of conventional thermo-chemical processes are accompanied by some distortion in shape or form.

In view of this situation, it has been traditional and, in fact, necessary when machining prior to heat treatment to leave a certain amount of stock to allow for post-heat treatment grinding in order to achieve the required final dimensions.

To a certain extent, the ferritic processes overcome some of the problems but the resultant heat treated condition whilst being resistant to 'scuffing' leaves a great deal to be desired when

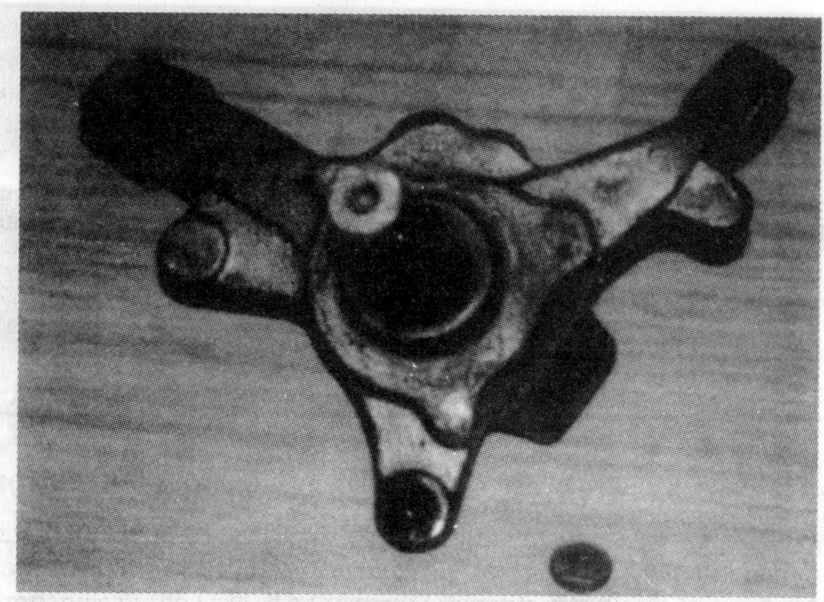

Fig. 13.6. Automobile forging produced on Zeta SA treated forging die

applied in an environment which requires excellent wear resistance, corrosion resistance and also indentation resistance.

These problems have been overcome with a portfolio of ferritic/austenitic and austenitic thermo-chemical heat treatment processes.

Each process correctly applied can overcome many long-standing manufacturing and service life problems.

The selection of the process to be applied is dependent upon the section thickness of material and the particular environment in which the component is required to operate.

AUSTENITIC PROCESSES

Alpha Plus Process

The Alpha Plus process is an ultra-low temperature process but is still carried out within the austenitic range.

Total case depth from 125 μm to 250 μm can be obtained depending upon the choice, which is superimposed with a nonmetallic compound layer of 25–50 microns in depth (Figs. 13.7 and 13.8).

The Alpha Plus process imparts the wear resistance expected of case hardened components, enhanced by the presence of the compound layer which, in itself, has a remarkable combination of properties, *viz.* wear resistance, anti-seizure capability, and corrosion resistance.

Movement during processing by virtue of the low process temperature is considerably restricted and is therefore particularly applicable to thin section pressings such as clutch plates, etc.

The development of Alpha Plus process has given the facility to process components to an improved metallurgical standard with associated distortion factors within acceptable limits thereby reducing scrap levels.

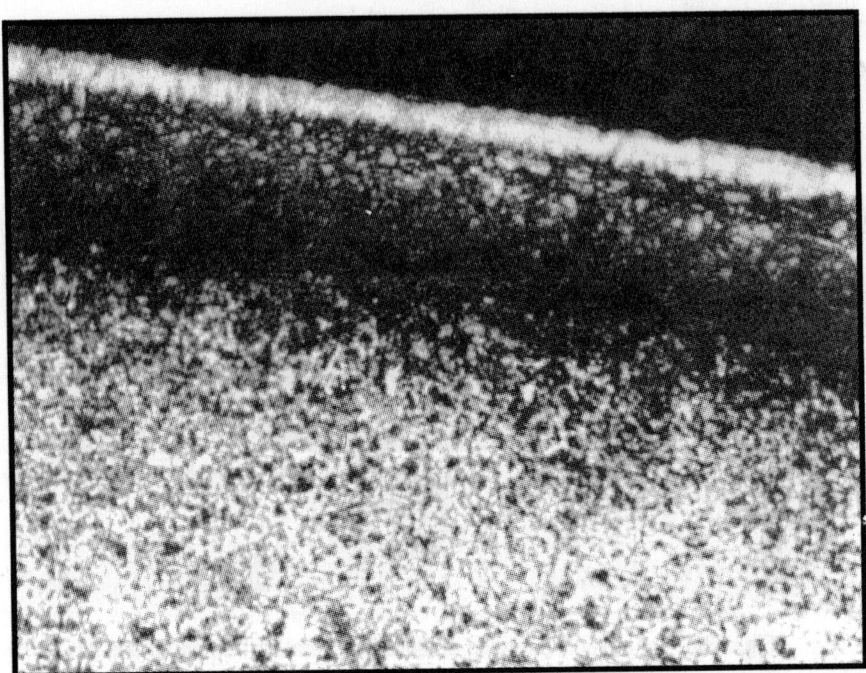

Fig. 13.7. Microstructure of Alpha plus processed mild steel

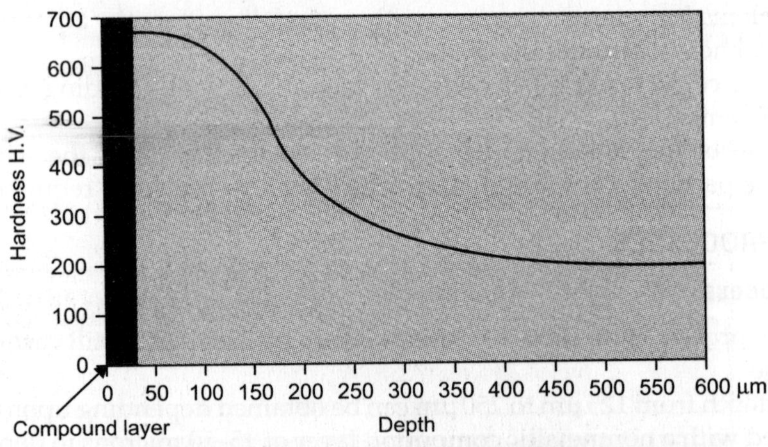

Fig. 13.8. Typical micro hardness traverse on Alpha plus processed mild steel

Also, the corrosion resistance produced by the process eliminates the necessity for post-heat treatment phosphating or plating which again manifests itself as a cost saving.

Alpha Plus processing is applicable to mild steels and low carbon steels.

The surface shelf life of processed components is far superior to that of traditional finishes.

Typical Applications of Alpha Plus Process

Clutch plates, valve parts, hydraulic cylinders, pump shafts, thin pressings, change speed gates, door hinges, thin walled bushes, small gears, handbrake ratchets, etc.

Beta Processes

The Beta family of processes are a series of cyclic processes aimed at sustaining the most aggressive wear conditions in a corrosive environment.

Beta 25 Process

The Beta 25 process is the basic Beta process and was developed for the processing of mild steel and low alloy materials for applications particularly in aggressive environments where conventional case hardened components and expensive materials had proven unsatisfactory, i.e. when subjected to a combination of aggressive wear and corrosion service conditions, e.g. brake parts for railcars.

Applications of Beta 25 Process

Beta 25 process can be applied to a variety of steels, e.g., mild steels, plain carbon steels and low alloy case hardening steels.

The basic Beta 25 process produces a case depth of 25 microns which is superimposed with a non-metallic compound layer of the order of 50 microns in depth (Figs. 13.9 and 13.10).

It should be noted that the microstructure of the Beta case is far removed from a standard case hardened structure, but imparts substantial indentation resistance with a combination of remarkable wear and corrosion resistance.

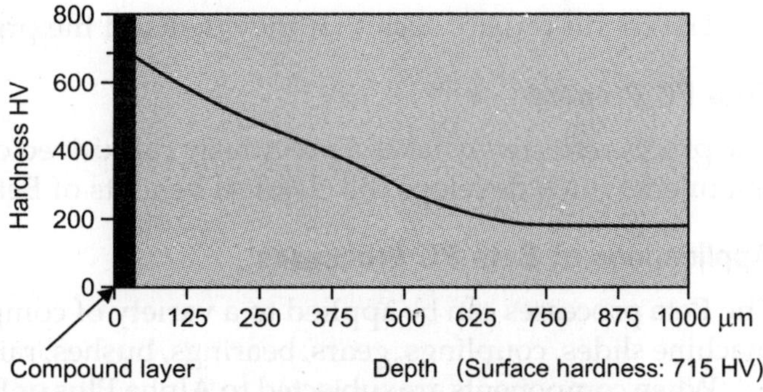

Fig. 13.9. Typical micro-hardness traverse on Beta 25 process treated mild steel (En. 32B)

Fig. 13.10. Microstructure of Beta processed mild steel

The phenomenal advantages of the process have enabled the mechanical/metallurgical engineer not only to solve long-standing service life problems but has also offered the facility in certain situations for the replacement of expensive steels with less costly steels whilst giving an improved performance.

Beta 40 Process

The Beta 40 process is a development of the Beta 25 process and produces a much deeper case depth of 1 mm but with the same depth of compound layer.

Obviously, the process is particularly applicable with heavy section components and for use in the more aggressive environments. Whilst most applications in this area can be covered by Beta 25 and Beta 40 processes some applications exist where extremely heavy wear conditions prevail.

The conditions are such that they demand the presence of extremely deep case depths.

Beta PC Process

The process effectively takes a previously carburized component and subjects it to a secondary treatment which develops the classical benefits of Beta processing.

Applications of Beta PC Processes

The Beta processes can be applied to a variety of components: pump components, chain pins, machine slides, couplings, gears, bearings, bushes, rail bogie pins and bushes, drill shafts, etc.

When components are subjected to Alpha Plus or Beta processes, areas can be masked with standard stop-off paint as in conventional processing.

FERRITIC/AUSTENITIC PROCESSES

Delta Process

The Delta process, again unique to Omega Technical Services, U.K. is the first ferritic/austenitic nitrocarburising process which has been developed and bridges the gap between ferritic nitrocarburising and austenitic nitrocarburising.

Material sections which cannot withstand austenitic nitrocarburising resultant case depths and yet have poor service life when ferritically nitrocarburised can now be processed satisfactorily solving a long-standing problem.

The compound layer/case ratio obtained from Delta processing is some 2/1 as against that obtained with austenitic nitrocarburising, i.e. 2/5.

Whilst the indentation resistance of Delta processed components is not as great as that obtained from austenitic nitrocarburising the deeper compound layer exhibits excellent scuff resistance and corrosion resistance.

The Delta processing temperature is in the ferritic/austenitic region which facilitates less shape distortion.

Applications of Delta Processes

- In pressed components (all pressings are required to be case hardened)
- In Garden Implement Industries (cutting blades, lawn mower blades)

The Delta SS series of processes were developed specifically to overcome the problems encountered with the nitriding of stainless steels.

Gaseous nitriding of stainless steels, in order to promote an increased level of wear resistance has always been subject to variation, particularly with austenitic stainless steels.

The problems have been overcome to a certain extent by the use of plasma-nitriding.

Plasma-nitriding plant is extremely capital intensive and, as such, limited facilities are available in the job-contract heat treatment industries.

The cost of plasma-nitriding is considerably higher than that of the gaseous processes and therefore good potential exists for a gaseous process which can demonstrate comparable reliability.

Omega Technical Services Limited, U.K., developed Delta SS processes to provide this facility (Figs. 13.11 and 13.12).

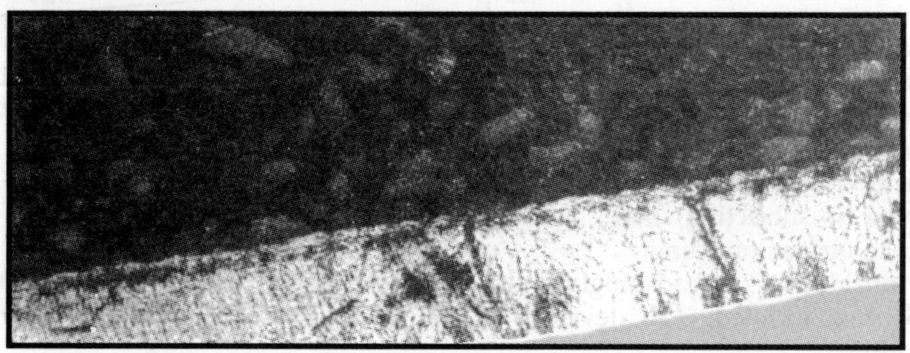

Fig. 13.11. Microstructure of Delta SSA processed austenitic stainless steel

The processing temperatures utilised for both Delta SS processes are sub-austenitic and consequently shape distortion can again be reduced to a low level.

It should be noted that some lowering of the untreated corrosion resistance is experienced by Delta processing as in all instances where stainless steels are surface treated.

The design engineer must weigh the advantages of increased wear resistance against the lowering of the corrosion resistance.

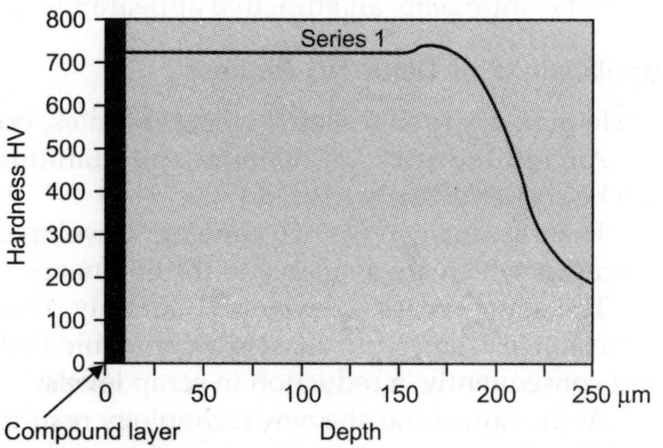

Fig. 13.12. Micro-hardness traverse on Delta SSM process treated of martensitic stainless steel

Further Applications of Delta Processes

- Dairy Engineering Industry (all stainless steel working parts requiring additional wear resistance)
- Medical Engineering (generally stainless steel parts requiring additional wear resistance)

- Food Engineering Industry (all stainless steel parts requiring additional wear resistance)
- Cutlery Industry (all forging dies and cutting tools)
- Instrument Engineering (all working parts requiring additional wear resistance)

Delta TiG Process

Delta TiG process was developed specifically for the treatment of titanium and titanium alloys.

Mechanical and metallurgical engineers wishing to capitalise on the excellent properties of titanium and its alloys have been searching for a cost-effective dependable process which promotes scuff resistance and wear resistance to components manufactured from these materials.

The improvement of scuff resistance and wear resistance has been achieved by subjecting such components to plasma nitriding.

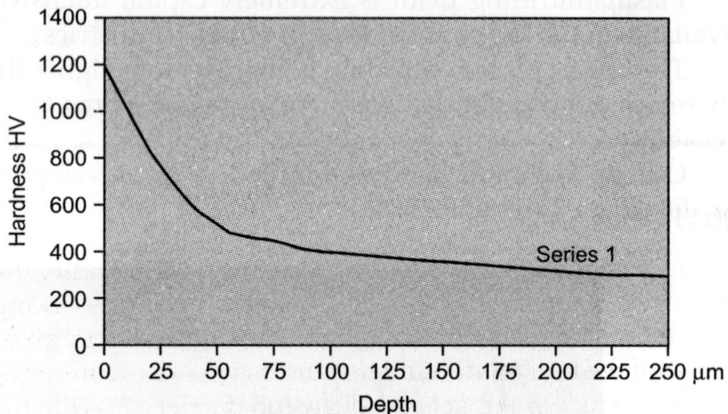

Fig. 13.13. Micro-hardness traverse on Delta TiG processed IMI 318 Titanium alloy

Plasma nitriding, as already highlighted, is an expensive process to carry out. Omega Technical Services Limited, U.K., developed Delta TiG as a direct and cheaper alternative to plasma nitriding.

The properties obtained closely match those obtained by plasma nitriding (Fig. 13.13).

It should be noted that the process imparts an associated gold colouration which gives the treated components an attractive appearance.

Applications of Delta TiG Process

Cycle Industry (pedal shafts, wheel spindles, bottom bracket shafts)

Aircraft Industry (all titanium and titanium alloy components requiring wear resistance can be cost-effectively treated.)

There are many types of technology developed over and above the traditional heat treatment processes which are available to the engineers.

The new processes developed not only offer improved properties in respect of wear and corrosion resistance, but are also accompanied by a greatly reduced distortion during treatment and consequently, a reduction in scrap levels.

At the same time, the new technology results in cost saving due to reduction in cycle times and the elimination of post-heat treatment cleaning, phosphating and plating, etc.

The new processes are nitrogen based and promote the conservation of energy by the elimination of direct energy consumption.

The processes described above are environment friendly when compared with traditional heat treatment and surface coating technologies.

14

Recent Developments in Surface Engineering

M. Farrow*

INTRODUCTION

Surface engineering embraces processes which modify the surfaces of engineering components to improve their 'in service performance', useful working lifetime, aesthetics, or economics of production. Engineering the surface is intended to give properties which are different from those of the bulk material. The purpose may be to minimise corrosion, reduce frictional energy losses, reduce wear, act as a diffusion barrier, provide thermal insulation, electrically insulate or simply to improve the aesthetic appearance.

With the increasing emphasis on improving the economics of manufacture and performance of engineering systems, whilst conserving energy and costly strategic materials the importance of exploiting the possibility of surface engineering cannot be overstressed.

MATERIAL SELECTION

Before a material is identified for a particular function three requirements must be satisfied:

(*i*) the bulk mechanical properties must be adequate for a useful operational life,
(*ii*) it must be feasible to fabricate the component economically in the quantities required,
(*iii*) the material must be capable of withstanding the environmental conditions during the design lifetime.

It is this third requirement which is of greater concern because wear, corrosion and other surface degradation processes can markedly curtail the life of an engineering component.

Some engineering components are placed in operational environments which provide practically no latitude for material selection, e.g., gas turbine blade materials must be selected on the basis of high temperature properties alone and chemical plant materials chosen on specific corrosion resistance in their intended environment. Turbine blade and ring materials are normally finished with high temperature erosion resistant surfaces and contacting areas with tribologically appropriate coatings. Similarly, chemical plant components are often clad with engineering steels, the cladding (weld or roll cladding) having the requisite corrosion resistance.

However, majority of the engineering components require a minimum design strength at low operating temperature with surfaces which have good wear properties, low friction coefficients and resistance to normal atmospheric corrosion.

This opens up the possibility of using the lowest cost substrate to satisfy the mechanical demands, and surfacing the components to meet the in-service environment demands.

---

*Consultant, Cemat Partners, Cheshire, U.K.

SURFACE ENGINEERING PROCESSES

There are many ways of treating metal surfaces to enhance the corrosion resistance and/or their tribological properties. These may be grouped into three broad categories (Fig. 14.1).

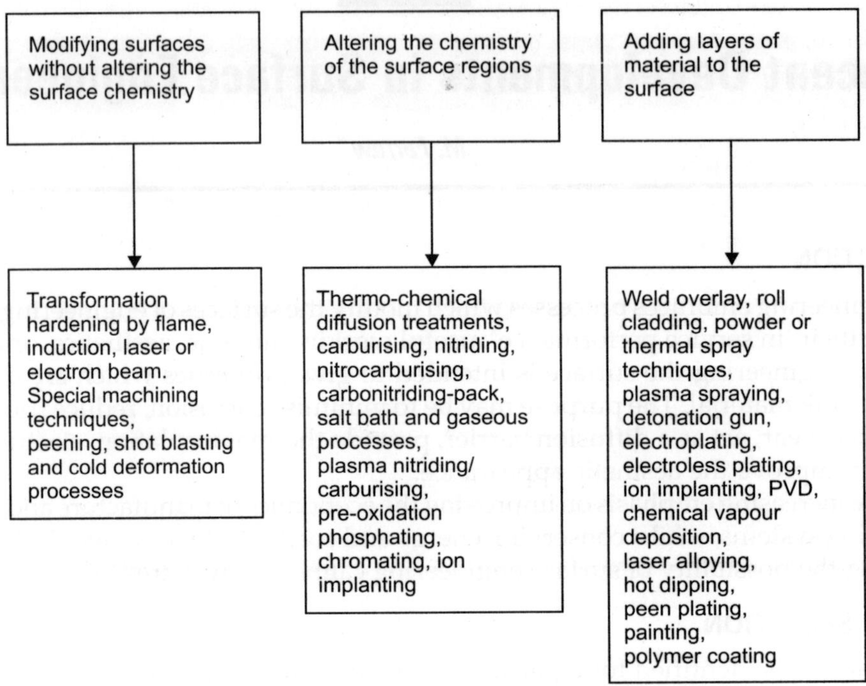

Fig. 14.1. Categories of surface engineering techniques

1. **Modifying the surface without altering the surface chemistry:** transformation hardening, cold deformation, machining and peening.
2. **Altering the surface chemistry** by chemical or thermo-chemical diffusion treatments—carburising, nitriding, anodising, ion implantation.
3. **Adding layers of material to the surface** by weld overlays, painting, metal spraying, electroplating, bonding, chemical vapour deposition, physical vapour deposition, laser glazing and roll cladding.

Techniques from any of these three categories may be used in combination with others, e.g., electroplated layers may be followed by a thermal diffusion treatment, metal spraying or anodising may be followed by sealing with a polymer, or a solid lubricant may be added to an already modified surface.

Category 1 Treatments

Transformation hardenable alloys

With carbon or low alloy steels and cast irons the option to harden using flame, induction, laser or electron beam techniques may be the most attractive. These processes can be fully automated and precisely controlled. The desired core properties can be developed by standard heat treatment practices and the surfaces hardened by rapidly heating them to 85°C and quenching. In some cases, it is prudent to follow the hardening cycle by a low temperature heat treatment to relieve the internal stresses. All of these processes may be usefully applied to selected areas of the surface without affecting the bulk properties. Depth of hardening limitations can occur with those processes that heat from the surface (surface liquation, oxidation and decarburisation). For this reason, induction processes have some advantage when treating massive components requiring deep hardening.

Laser hardening

Laser hardening can now compete in high volume production with other low cost processes, such as induction hardening. Using self-quench techniques it is possible to obtain case depths of 0.75 mm. Lasers are particularly useful for hardening relatively inaccessible areas. Another attribute is that they can be focused to a spot to harden a precisely defined area, e.g. inside cylinder liners. Lasers provide fast production, low distortion and avoid the necessity for post treatment operations, including machining.

Electron beam hardening techniques

Electron beam hardening techniques have similar attributes to lasers and may be more economical because both capital and operational costs are lower. The beam operates in a vacuum but the workpiece need only be at 60 m bar pressure. Area hardening is obtained by scanning the area on a raster: dwell time, beam energy and focus all being computer controlled. This process is used commercially for automotive components where induction hardening would produce too much distortion.

Cold working

Cold deformation by peening, shot blasting and other processes, which produce surface stored energy and compressive stresses, increase the hardness, wear resistance, fatigue properties and stress corrosion resistance. These processes are specialised and are only applicable to certain alloys and applications. However, the surface hardness and topography are altered, both of which have tribological advantages.

Category 2 Treatments

Thermo-chemical diffusion treatments

Thermo-chemical diffusion treatments mainly introduce interstitial elements, such as carbon, nitrogen and boron, or combinations of carbon and nitrogen into ferrous metal surfaces at elevated temperatures. The processes are not confined to interstitial diffusion, metallic substitutional elements or metalloids are used in processes such as chromising, aluminising and siliconising.

Interstitial element diffusion falls into broad groups, those carried out at low temperature, i.e. within the ferritic range; or high temperature treatments in the austenitic range. Ferritic processes include gas nitriding, typically 525°C, plasma nitriding 400–600°C and nitrocarburising processes approximately 500°C. For ferritic nitrocarburising processes many different treatment media may be employed including salt baths (cyanides or non-toxic mixtures) endothermic ammonia gas mixtures and methanol propane/ammonia/oxygen mixtures.

The austenitic treatments broadly include carburising employing solid (pack), liquid (salt bath) or gaseous media; boronising and carbonitriding. Thermo-chemical treatments involving substitutional elements, such as chromium (chromising) or aluminium (aluminising), may be pack, salt bath or vapour processes. The surface engineered products are often used for high temperature service utilising nickel based or superalloy substrates for use in gas turbines.

Electroplating and thermal diffusion

Electroplating and thermal diffusion treatments, when used in combination, are included in this category. One process involves the electrolytic deposition of tin onto ferrous materials. This is followed by a diffusion treatment at 400–600°C to form Fe-Su compounds which resist scuffing and give some corrosion resistance to the components. Bronze coatings may be developed in a similar way to add a bearing surface to a steel substrate. Furthermore, treatments are available for non-ferrous substrates, such as titanium, copper and aluminium base alloys to provide anti-scuffing layers and enhance the tribological compatibility with the mating surface.

Oxide coatings

Oxide coatings on the surface of components can produce significant tribological advantages, particularly in preventing adhesive wear. When oil is present they prevent scuffing, adhesive wear and metal transfer. On ferrous substrates, chemical conversion layers may be produced by immersion in caustic nitrate solutions. This type of process is applied to needle or roller bearings, gears and piston-rings. Similar coatings can be developed by thermal exposure at 300–600°C to produce an oxide film. Steam tempering or autoclaving is applied to high speed steel drills and zirconium alloy components for this purpose.

Sulphur treatments

Sulphur treatments incorporate sulphur into the surface of ferrous components. Sulphur, because of its low melting point and some sulphides because of their crystal structures, have good lubricating properties. These processes are used for anti-scuffing purposes on cylinder liners, gears, constant velocity joints, heavy duty rear axle components, textile machinery parts, etc.

Anodising treatments

Anodising treatments for aluminium, magnesium, zinc and titanium alloys produce oxide layers which prevent adhesive wear and are generally harder than the substrate. It is also interesting to note that anodising may be followed by treatments to seal the surface and improve the corrosion resistance, or incorporate solid lubricants into the surface to lower friction and reduce wear-rates.

Phosphating

Phosphating, BS 3189, 1973 gives five types of coatings that might be used. The two heaviest give maximum corrosion protection, particularly under grease or oil, but as their cohesive strength is poor these are unsuitable as a base for painting. All phosphate coatings absorb oil and grease, thereby assisting 'running in' by preventing adhesive wear and fretting.

The precise details of phosphating processes are generally proprietary but are based on dilute phosphoric acid solutions of iron, zinc and manganese phosphates. Accelerators are added to shorten the process times to just a few minutes at 40–70°C. The simplest phosphate coatings consist of grey or black crystals of $Fe_3(P_2O_4)_2$ and some $FePO_4$. Zinc and manganese produce more complex layers which absorb lubricant more readily. They are effective in reducing galling, pickup and scuffing.

Ion implantation

In this process, atoms of gaseous or metallic elements are ionised and sent to a high vacuum chamber where they are accelerated through a mass separator. Selected ions are further accelerated and implanted into the target component. The implanted species occupy interstitial sites and distort the lattice. It is a low temperature process, typically 150°C for small items and less for larger components. The depth of effect is very shallow, 2×10^{-4} mm but the surface properties such as wear resistance, friction and oxidation/corrosion resistance can be considerably enhanced. This process has been used to enhance the performance of forming tools for plastics, press tools and some surgical implants.

Category 3 Treatments

There are numerous processes which involve coating with a layer of material, not necessarily metallic, to meet the requirements of specific service environments. Such processes are cost effective and it is possible to overcome most corrosion and tribological problems in this way.

Weld or roll cladding

Weld or roll cladding usually involves relatively thick layers. Weld cladding can be used to good effect where abrasive wear is a problem, such as coating earth mover equipment bucket teeth, tank tracks and mineral handling equipment. Roll cladding is usually associated with corrosive or mild erosive wear problems, typically those encountered in the chemical, wood pulp, paper and food process industries. The two processes are, to a certain degree, complementary. Hard, abrasion resistant materials that are impractical to fabricate by conventional techniques are best deposited by welding. Whilst the more corrosion resistant materials are more amenable to cladding or forming.

Metal spraying

Metal spraying involves heating metal, ceramic or mixtures of metal and ceramic powders to a semi-molten state and depositing them at high velocities onto the components. These processes can be divided into flame gun, arc, plasma-arc and detonation gun techniques. Difficulties with these processes are accessing the coating to substrate bond, integrity, porosity and general quality of the coating on a production basis. For high integrity coatings the application of hot isostatic pressing (HIP) after coating has been found to seal the porosity and improve the bond to substrate quality.

Electroplating

Electroplating—over 30 metals can readily be deposited from aqueous solutions. Alkali/alkaline earth metals and refractory metals, such as tungsten, molybdenum and vanadium cannot be deposited from aqueous solutions. There is a tendency to think that electrolytic deposits are mainly for corrosion resistance, decorative or electronic/electrical usage but there are many engineering and tribological applications of electroplating. Hard or soft deposits are used depending on the particular function required. Hard chromium plating is ideal for resisting abrasive wear, pickup and for corrosion/abrasion resistance. Porous or intentionally cracked chromium deposits for oil retention, as in automotive cylinder liners, precision bearing sleeves and piston rings. Soft deposits such as tin are used to facilitate 'running in', prevent fretting and to impart corrosion resistance. Deposits of silver, lead, cadmium, tin and antimony are used in heavy duty sleeve bearings; typically in aircraft power units. Nickel can be deposited from a wide range of solutions and finds application to minimise abrasive wear of sliding mechanisms.

Care should be taken when selecting nickel, particularly with respect to its potential mating surface because it has a tendency to gall. Nickel is a good base coating for chromium and because the shock resistance of chromium is poor it is prudent to make the bulk of the layer nickel and give a relatively thin top layer of chromium.

Electroless plating

Electroless plating, the autocatalytic deposition of nickel phosphorous and nickel-boron may have useful corrosion and tribological applications. In the case of Ni-P deposits a hardness of about 500 DPH (Diamond Pyramid Hardness) is obtained but thermal ageing at temperature around 400°C can develop hardness values in excess of 1000 DPH. These values are not retained at high temperatures although Ni-B deposits are superior in this respect. Excellent results have been obtained with Ni-B on glass container manufacturing moulds.

Composite electroplated and electroless plated deposits

Composite electroplated and electroless plated deposits involve the production of electroless and electroplated metals into which micron sized dispersions of non-metallics are incorporated. Composite coatings of electroplated nickel containing silicon carbide particles exhibit superior abrasive wear resistance to hard chromium plating in some applications. Electrolytically deposited cobalt incorporating chromium carbides has been successfully used in both dry and lubricated conditions at 800°C. Incorporating 7–8 μm sized particles of PTFE (Poly-Tetra Fluoro-Ethylene) as a solid lubricant in nickel coatings produces low friction self lubricating surfaces. These are finding extensive use in the offshore oil and automotive industries.

Chemical vapour deposition (CVD)

Chemical vapour deposition (CVD) involves the dissociation of metal compound vapours at temperatures in excess of 850 °C to produce thin, diffusion bonded, adherent coatings of metal carbides, nitrides, carbonitrides and oxides; typically TiN, TiC, Ti (CN) and Al_2O_3. The process is amenable to producing duplex and layered coatings which can help to achieve good compatibility with the substrate, incorporate diffusion barriers and obtain optimum properties on the outside surface of the coating system. CVD has been used for more than 30 years as a coating on carbide tool tips (indexable/inserts) and on selected tribological items. The high temperature of the coating process limits the choice of substrate to high temperature materials and items with simple geometrics that do not distort excessively. Only selected ferrous items can be treated, e.g., carbides with cobalt binders or high speed steel tools with simple shapes (the latter permitting them to be re-heat treated after deposition). However, techniques for plasma assisted chemical vapour deposition (PACVD) are being developed which will permit coatings to be deposited at temperatures well below the tempering temperatures for high speed steel i.e. < 500°C.

Physical vapour deposition (PVD)

Physical vapour deposition (PVD) is becoming increasingly important for small engineering components. PVD embraces evaporative deposition, sputtering and ion-plating in inert or reactive environments. Process temperatures are relatively low 40–400°C, thus minimising distortion and preserving the heat treated state of the substrate.

Reactive plating

Reactive plating takes place in an inert gas to which is added a small partial pressure of a reactive gas to supply carbon or nitrogen. The metallic species is added to the system by resistance heating, arc or electron beam evaporation or sputtering from a solid target. Nitrides of titanium, zirconium, hafnium or chromium and other metals have been deposited onto metallic components to provide 3–5 μm of hard layers (> 3000 DPH) of inert, low friction coefficient compounds.

These ceramic layers enhance the performance of cutting tools and have considerable potential for many other small items. However, the process is limited to relatively thin layers. For heavily loaded contacts the supporting substrate must have a relatively high hardness to contain the Herzian stresses.

Sputter ion plating techniques are also used to deposit solid lubricants MoS_2 PTFE and lead onto bearing surfaces for service in vacuum systems and space satellites. Corrosion protection layers and coating compounds for high temperature tribological service in gas turbines are also effectively deposited by PVD techniques.

Painting

Painting to protect a substrate against corrosion and improving the aesthetic appearance is probably the best known surface modification process. There have been considerable advances

Table 14.1. Methods of coating technology

Coating technique	*Coating type*			
	Metal	*Metallic*	*Paint*	*Organic*
Electro deposition (Electro-plating)	•			
Electroless plating	•			
Electrophoretic plating or coating	•	•	•	
Hot dipping	•		•	
Spraying	•	•	•	•
Cladding	•	•	•	•
Vacuum/Vapour deposition	•			•
Vapour dissociation	•			
Cementation and diffusion coating	•	•		
Fusion coating		•		
Roller coating				•
Conversion coating		•		

in paints, application techniques and pre-coating/painting treatments. Surface preparation and corrosion protection methods, such as phosphating, have brought painting into the range of engineering coatings. Table 14.1 shows how painting and organic coatings form part of the spectrum. Organic coatings deposited on metal parts by spraying, brush application or dipping are replacing electroplated deposits on some automotive parts.

Painting, dipping or spraying with organic resins and polymeric materials, to which metallic ceramic or solid lubricant compounds are added is providing for both corrosion and tribological requirements.

One process consists of zinc flakes bonded with zinc chromate and a proprietary organic material. This process provides excellent surface protection and is widely used in the automotive industry for fasteners, springs, sintered parts and items for steering gears.

SELECTING THE APPROPRIATE SURFACE ENGINEERING PROCESS

With due consideration to the service environment, particularly wear and corrosion, at the design stage, substrate selection is relatively easy. Ideally, the lowest cost component which will satisfy the engineering demands is desired. However, components with demanding duties (functional requirements) will necessitate additional expenditure on either upgrading the substrate or surface engineering a cheaper substrate.

The chosen surface modification process must be of adequate thickness to withstand the contact stresses throughout its operating lifetime. Processes, which produce surface layers of only a few microns, are only suitable for low loads or mildly corrosive conditions. Figure 14.2 indicates that the coating thickness ranges for some surface engineering processes. An indication of relative costs are given in Figure 14.3. That also shows that high technology coatings need to be selected with care to justify them on economic grounds.

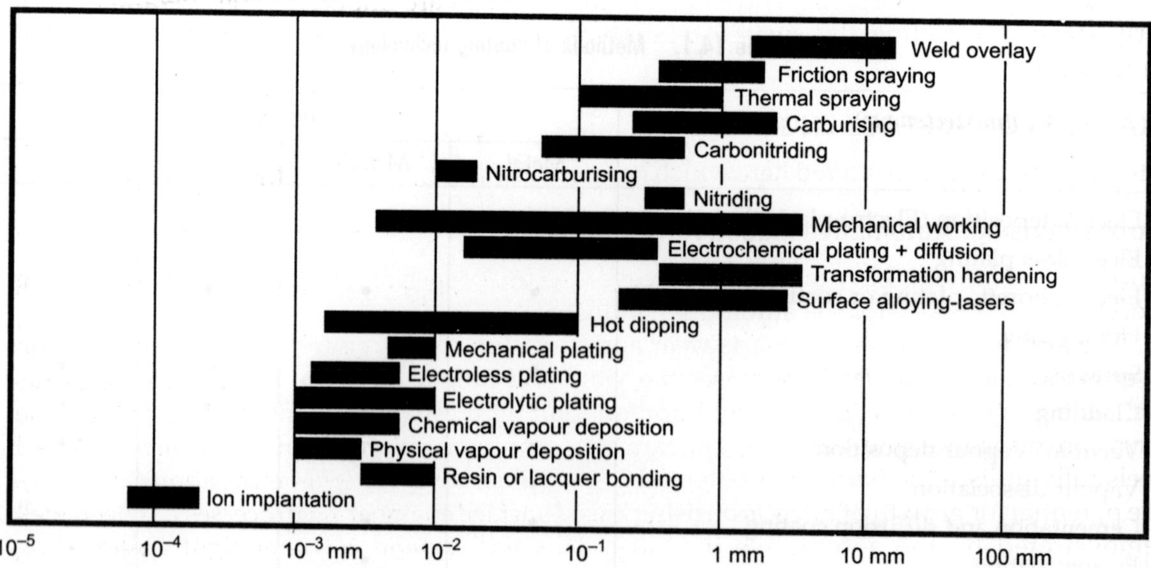

Fig. 14.2. Approximate thickness of treatment effect

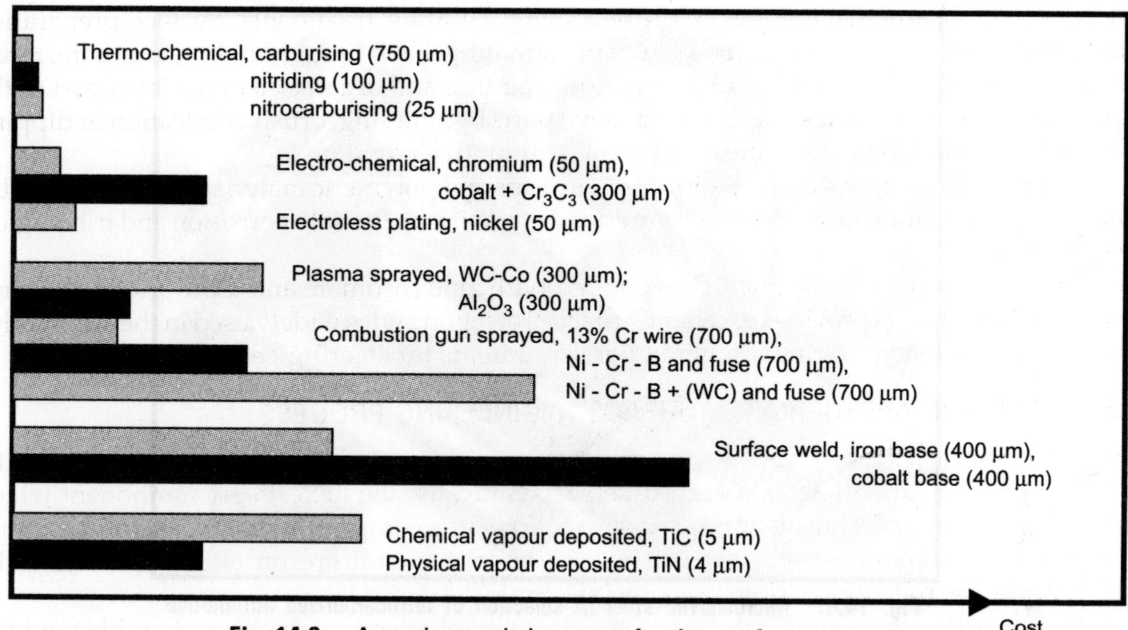

Fig. 14.3. Approximate relative costs of various surface treatments

Thermo-chemical treatments and high temperature processes often cause distortions or dimensional changes in items of complex geometry. This should be appreciated and evaluated. Similarly, the condition of the surface after treatment must be considered, particularly if a further machining operation is required.

The component mass, size and numbers to be treated are important considerations at an early stage in selecting the surface treatment and for considering automated processing (automation).

EXAMPLES OF SURFACE ENGINEERING

A few examples are considered here which has potential for much wider industrial applications.

Nitrocarburising: Automotive industry

Process developments have markedly increased the use of specialised gaseous nitrocarburising processes, particularly for the automotive industry.

The nitride surface layer imparts wear and corrosion resistance and the nitrogen diffusion layer below the nitride layer increases the yield and fatigue strength. Recent developments such as post nitriding oxidation and impregnation with organic sealants have increased the corrosion resistance and aesthetic appearance. Also fabrications from sheet micro-alloyed steels fully utilises the hardening potential of the substrate by internal nitridation (Fig. 14.4). The potential for growth of nitrocarburising based surface engineering processes has been well demonstrated by the automotive industry. This will extend to other light engineering manufacturing. Improved products of lower mass will be produced cost effectively for a wide range of industries in due course (Tables 14.2 and 14.3).

Fig. 14.4. Microalloyed steel (A selection of nitrocarburised automotive components)

Table 14.2. Comparison of surface treatments

Surface Treatment Method	*Strength Enhancement (Low alloy steel)*		*Surface Hardness HV*	*Salt Spray Corrosion Resistance ASTM B117 (Hours)*	*Treatment Cost per unit area as an index to Barrel Zinc plating*
	Yield strength	*Fatigue strength*			
Carburising	3–6 times	Up to 3 times	900	< 20	1–2
Nitrocarburising (*)	3–4 times	Up to 4 times	> 1000	< 20	2–3
Nitrocarburising with post-oxidation	3–4 times	Up to 4 times	> 1000	> 250	2–3
Hard chromium plating	–	–	900–1000	120	18–36
Electroless nickel plating	–	–	550–1000	100–250	22–28
Nickel electroplating	–	–	120–400	> 250	14–20
Zinc electroplating (10 μm) + passive	–	–	Very soft		
Rack				150–200	2
Barrel				150	1

*Note: relatively low cost of nitrocarburising

Table 14.3. Design property requirements for a selection of applications

Properties	*Applications*														
	Torque plates	*Seat slides*	*Brake stop control system*	*Bumper beams*	*Helical gears*	*Bearing bushes*	*Transmission gears*	*Door locks*	*Windscreen wipers*	*Gas spring pistons*	*Suspension strut pistons*	*Alternator cooling fan*	*Fan motor*	*Horn*	*Sunroof actuator*
Wear resistance	•	•	•		•	•	•	•	•	•	•		•	•	•
Indentation resistance		•	•				•				•				
Yield strength		•		•			•	•	•		•	•	•	•	•
Fatigue strength		•			•				•		•	•		•	
Bearing characteristics						•			•				•		•
Corrosion resistance		•	•	•				•	•	•	•	•	•	•	•
Surface finish	•		•			•			•	•	•		•		
Aesthetic finish		•						•	•	•		•	•	•	
Dimension control	•	•	•	•	•	•	•	•	•	•	•	•	•	•	•

Ion-plated titanium nitride: Cutting tool industry

High speed steel cutting tools, which are ion-plated with approximately 3 mm of titanium nitride, have much longer lives between sharpening (grinding), allow higher metal removal rates and leave an improved surface finish compared with uncoated tools. A cutting tool manufacturer has redesigned one of their ranges of twist drills to maximise on the benefits afforded by TiN coating. The design allows much higher drilling and metal removal rates. This example demonstrates what can be achieved by designing to fully utilise the properties of surface engineered surfaces. Figure 14.5 shows the comparative performance of 2 standard drills and one TiN coated drill of the same design.

Ion implanting: Hip prostheses

The femoral component for total hip replacement made from 6 per cent Al, 4 per cent V. Titanium alloy operates against an ultra high molecular weight polyethylene acetabular cup. When the titanium alloy part is ion-implanted with nitrogen it exhibits a wear reduction of 500 to 1000 times when compared with an untreated part; the new system could provide protection for up to 30 years.

Sputter ion plating of solid lubricants: Vacuum systems

The lubrication of rolling element bearings in space satellites and other vacuum systems must be with low vapour pressure oils or solid lubricants. The former have very limited application and sputtered lead, PTFE or molybdenum disulphide are most effective.

Sputtering techniques have been developed to deposit lubricant layers of 0.1 μm thickness. Extremely low friction coefficients < 0.01 have been obtained with sputtered MoS_2. Deposits on the raceways of rolling element bearings have demonstrated very effective lubrication for long durations in space satellites.

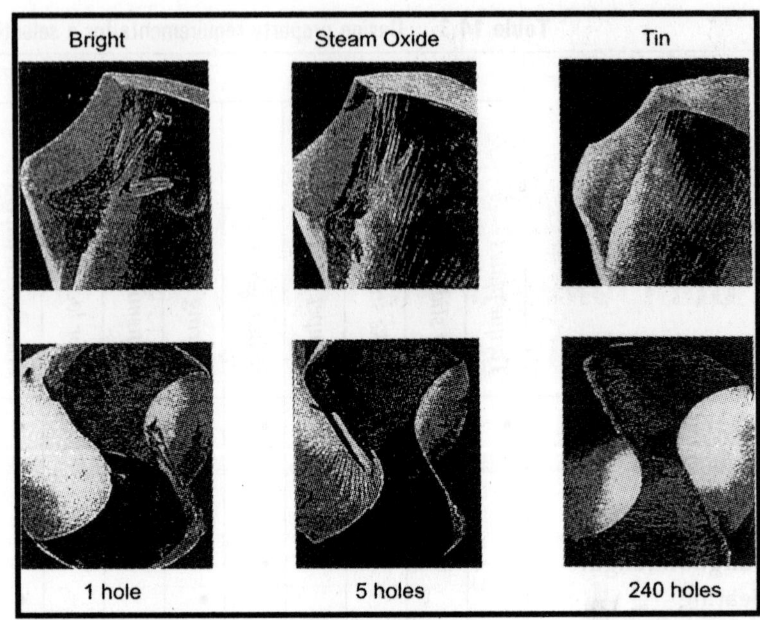

Fig. 14.5. **Screech tested drills. Holes are continuously drilled through a 25 mm thick mild steel plate at standard speed and feed rates. The test is stopped when the drill screeches. Photographs show the drill condition after the test**

Old and well established surface modification processes have considerable merit for improving the corrosion and tribological properties of engineering components. In several areas it has been demonstrated that the new surface treatment technologies arising from the use of lasers, electron beams, and plasmas complement those already in existence.

New technology materials, such as polymers, ceramics, solid lubricants and resin bonded composites, can, when applied as coatings to component surfaces, enhance the wear or corrosion resistance and in some cases reduce friction.

Clearly the cross fertilisation of established techniques with modern application methods and materials will have significant effects in enhancing the performance of engineering surfaces.

Surface engineering can also produce components with high operating efficiencies which are unobtainable by other techniques.

There is considerable scope for industry to utilise established and modern surface engineering techniques to get improvements in their manufacturing techniques.

15

Advances in Surface Coatings for Enhanced Life of Components and Improved Products Quality Under High Pressure and Temperature Environments

*Dr. K. Anand**

INTRODUCTION

The engineering industry today is increasingly being pushed to operate under most severe conditions wherein the components are exposed to high speeds, temperatures, pressures, and often under corrosive environments. Surface degradation of components under the above conditions have a direct bearing on the useable life of the components and quality of the products being produced.

Surface engineering has emerged as one of the most important tools in enhancing component life. Quality products can be mass produced without frequent changes or wear out of components in machinery, if surface is treated appropriately.

A variety of surface engineering techniques are available with specific applications in steel, paper, textiles, power generation, chemical engineering, manufacturing, aircraft industries, etc. Examples of such coating materials include carbides, ceramics, nitrides, fluoropolymers, etc. Such coating materials are applied by techniques, such as thermal sprayed deposition processes, PVD (Physical Vapour Deposition) and CVD (Chemical Vapour Deposition).

Surface engineering techniques are playing an increasingly important role in the industry in two major areas.

- **Improved product performance:** Examples include the use of wear resistant or oxidation resistant and high temperature resistant coatings in aircraft engines, advanced process pumps and compressors and automobile engines. The adoption of coatings as an integral part of a new product line is driven by the need to produce better, more reliable machines that deliver better product performance in terms of speed, power and efficiency.
- **Improved quality and productivity in a manufacturing process:** In today's competitive environment, the industry is constantly trying to refine its manufacturing process to improve throughput, reduce defects and produce products with better quality. Total Productivity Management (TPM) is becoming an important goal for competitive industries. Surface engineering is emerging as one of the tools towards the attainment of such a goal.

**Manager, Business Development, Surface Tehnologies, Praxair India, Bangalore, India.*

Modern surface engineering techniques and examples of applicability of such techniques towards delivering better product performance or improved product quality in a manufacturing type environment are presented here.

SURFACE ENGINEERING—AN OVERVIEW

Surface engineering essentially involves altering the surface properties through modifying the existing surface (and immediate sub-surface) structure through phase transformations or introducing of additional elements into the surface such as N (Nitrogen), B (Boron) and C (Carbon), or adding a layer of a new material such as a carbide, nitride or ceramic coating as an overlay to provide unique wear, corrosion and temperature capabilities. The surface engineering techniques overcome the limitation of bulk materials which are engineered towards meeting properties such as mechanical strength, toughness and producibility (manufacturability) but are otherwise unable to meet wear, corrosion or high temperature requirements. A surface engineered component is thus a marriage of optimized bulk material properties with enhanced surface properties to meet specific operating conditions. Fig. 15.1 provides a broad spectrum of surface engineering techniques that are available and the materials that can be applied.

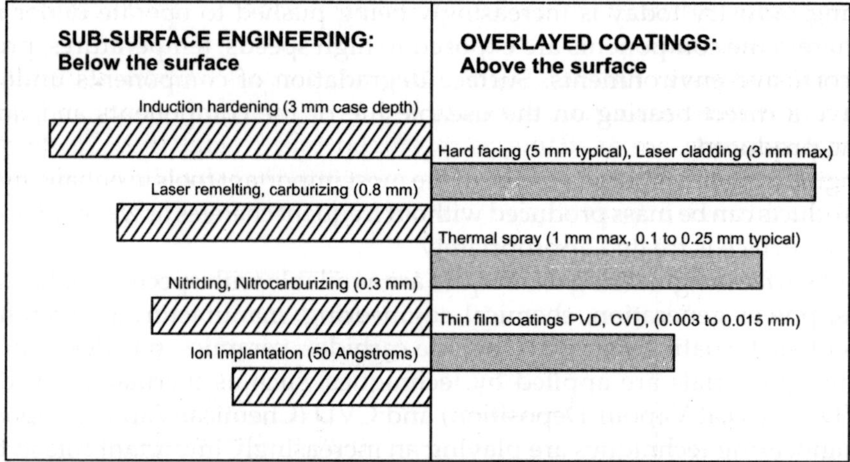

Fig. 15.1. Overview of the major surface engineering processes

1. **Ion Implantation:** In this process, ions of elements such as Nitrogen or Boron are accelerated towards the components and forced to embed beneath the surface. Typically the ions penetrate by about 5 nm (50 Angstroms) below the surface. This results in local surface hardening and induce compressive residual stresses several tens of microns below the surface. The wear, fatigue and corrosion properties are greatly improved. Typical applications include orthopaedic implants and bearings. Ion implantation has emerged from a laboratory level process to an engineering application.
2. **Laser Hardening:** This process is similar to induction hardening in principle; however, because of the highly localized heat input from a laser beam, it is possible to exercise much better control over the region that is undergoing martensitic transformation. Thus, problems such as distortion or cracking, are reduced substantially. Laser hardening

is commercially applied in a few automobile applications with great degree of success; examples include laser hardening of cam shafts, etc.

3. **Thin Film Coatings (PVD/CVD Processes):** PVD TiN (Titanium Nitride) coatings and CVD multilayered coatings for carbide inserts are relatively well established in India. These coatings are typically 3 to 15 microns in thickness and have a hardness in the range of 1800 to 2300 HV. Because of its high hardness and relative inertness the coating imparts wear resistance and allows higher cutting speeds. Over the last decade or so a variety of other coating compositions have been developed and gradually commercialized especially for cutting tool applications.
4. **TiALlN:** An improvement over TiN because it can withstand higher temperature variations. The better high temperature properties are attributed to the *in situ* formation of Aluminium Oxide. Useful for higher speeds compared to TiN.
5. **TiCN:** High hardness and better heat transfer properties compared to TiN. Particularly useful for drill bits.
6. **CrN:** Excellent antigalling and antisticking properties. Particularly useful for machining of Aluminium based alloys. CrN hardness is 1800 HV which is slightly lower than TiN.
7. **Sputtered Molydisulfide:** Used in conjunction with a hard thin film coating as the backing layer. Particularly useful in punching tools as it reduces the coefficient of friction substantially.
8. **Thermal Spraying:** This is a highly versatile process wherein a metallic, carbide or even a polymeric material can be applied as a coating. A variety of processes under thermal spraying are available as explained below (Fig. 15.2). In thermal spraying process the material to be coated is fed typically in the form of a powder and sometimes as a wire through a spray device/nozzle.

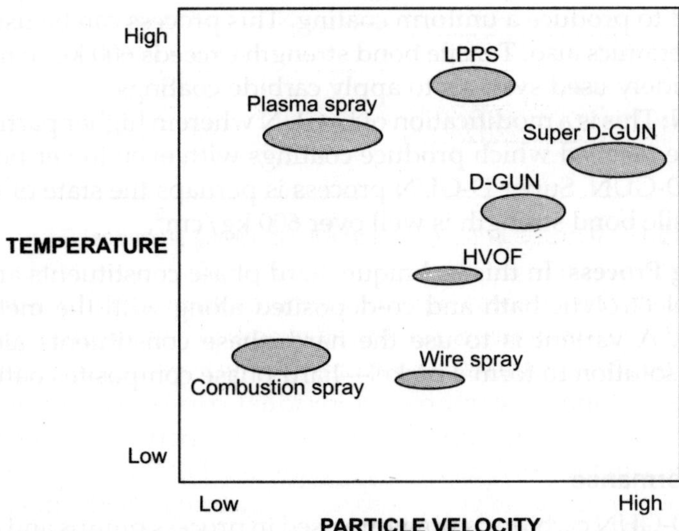

Fig. 15.2. Temperature–Particle velocity diagram of the various sprayed deposition processes. The velocities refer to that of the particles and not the gas flow

(a) **Combustion Spray:** The simplest thermal spray device is a combustion spray which uses a low velocity oxyacetylene flame to melt and deposit the particles; the particle velocity is very low and the adhesion strength is poor.

(b) **Plasma Spray Process:** In this process, a combination of gases such as nitrogen or argon with hydrogen is fed through a gun and is ionized as it passes between a tungsten cathode and a water cooled copper anode (nozzle). The ionized gases recombine with the electrons in the plasma generating tremendous heat. Particles fed either externally or internally (into the gun) to the plasma plume melt and get accelerated to moderate speeds of the order of 400 m/s. The result is a coating with moderate adhesion strength (150 to 450 kg/cm^2, typical value). This system is ideal for ceramics and is also used extensively to deposit metallic compositions and carbide compositions.

(c) **HVOF:** This is an improved combustion system using hydrogen, propylene or acetylene as the fuel gas. The gas is fed at higher pressures and burnt with oxygen. A supersonic velocity combustion gas plume is generated through which particles to be coated are fed. The particles melt or partially melt and deposit on to the surface. This system cannot melt ceramics easily. Typical particle velocities are 650 m/s. Coatings with < 2% porosity and adhesion strength that exceeds 600 kg/cm^2 are produced. This system is most suited to deposit carbide coatings.

(d) **D-GUN and Super D-Gun:** (These are proprietary processes using specially designed gun, and equipment system). In D-GUN technique, particles to be coated are fed into the gun along with a metered amount of fuel—Oxygen mixture. This mixture is ignited which results in a detonation—releasing a high temperature or high velocity wave. Particles get accelerated to > 750 m/s and deposit to a porosity that is < 1%. This cycle of particle ingestion and detonation is repeated several times a second to produce a uniform coating. This process can be used to deposit low melting ceramics also. Tensile bond strength exceeds 600 kg/cm^2. This is one of the most widely used systems to apply carbide coatings.

(e) **Super D-GUN:** This is a modification of D-GUN wherein higher particle velocities (>900 m/s) are reached which produce coatings with even lower porosity levels compared to D-GUN. Super D-GUN process is perhaps the state of the art in the industry. Tensile bond strength is well over 600 kg/cm^2.

9. **Composite Plating Process:** In this technique, hard phase constituents are colloidally suspended in an electrolytic bath and co-deposited along with the metal to form a composite coating. A variant is to use the hard phase constituents along with an electroless plating solution to form a nickel—hard phase composite coating.

APPLICATIONS

To Enhance Product Performance

- D-GUN and Super D-GUN carbide coatings are used in process pumps and compressors to reduce component wear and therefore leakage of process gases thus greatly enhancing the safety and reliability of the equipment.
- D-GUN and HVOF carbide coated gate valves for oil field environment; the coatings allow the component to withstand high temperatures and pressures in the presence of

abrasive media without permitting any wear. As a result, the surface finish of the sealing surfaces is maintained thereby eliminating leakage.

- D-GUN and Super D-GUN coatings in aircraft engines reduce component wear and erosion without adversely affecting the fatigue life of components; the process ensures favourable residual stresses and the compositions selected ensure the right balance between hardness and wear resistance versus toughness.
- Thermal barrier coatings are used in high pressure turbines to reduce base material temperature. This is instrumental in prolonging life of such components because creep and oxidation are retarded and the threshold limits for stress rupture are pushed higher.
- CrN coated piston rings reduce ring and cylinder liner wear.

To Enhance Productivity and Quality

- PVD and CVD thin film coatings on tool inserts permit the use of greater machining speeds and greater depth of cuts while improving tool life. Because of seizure resistance, the surface finish of the machined surface may also show improvements.
- Super D-GUN coated tungsten carbide coatings allow machine calendar rolls in paper mills to operate for a longer period without the need for regrinding. This eliminates periodic mill shut-downs and ensures constant paper quality and better printability.
- Ceramic coated textile machinery components prevent abrasion of the components by fillers and pigments present in the fibre being processed while maintaining a surface that is inherently fibre friendly. The use of D-GUN coatings allows the generation of special surface textures that reduce fibre-roll friction for specific processing conditions.
- The use of fluoropolymer topped hard coatings in tyre moulds ensure easy removal of the tyre and eliminate tearing of the tyre because of adhesion.
- D-GUN coated cermet coatings on furnace rolls eliminate material pick up between the cold rolled coil and the roll during annealing operation. This eliminates one of the most tricky problems in continuous annealing and galvanizing lines during the manufacture of cold rolled coils.

A wide variety of well established surface treatments are now available to improve product performance and to improve productivity and quality in a manufacturing process. These treatments have proven to be extremely cost effective value engineering tools which enable a manufacturing operation to run more productively with lower defect rates. The unique properties imparted to a surface through coatings can ensure better wear, corrosion and oxidation resistance to the components and enable them to perform better and more reliably in service.

16

Materials Planning and Logistics in Automotive Industry

Jason Shang *

TRENDS IN AUTOMOTIVE INDUSTRY

Competition in the automotive industry is severe and only high quality, low cost and speedy players can survive in this battle.

In the future, the competition will become more and more tough and more companies will be merged. There may be only 4 or 5 major players globally and some small regional players may survive.

THE ROLE OF MATERIALS PLANNING AND LOGISTICS

- To supply right material in right quantity to the production line to ensure "build to schedule" and to prevent any line stoppage because of material shortage.
- Only a satisfied customer can ensure profit for car makers. Automotive manufacturing companies should ensure that customers get their vehicle within acceptable waiting time at lowest transportation cost. Since customers get more and more choices, no one will wait, the lead time from receiving order to delivery is going to be very crucial for any company's survival.

Challenges:

Basic functions and challenges for automotive manufacturing industries are:

- Stable forecast of material demand. Various companies use different strategies to maintain their stability. Most commonly adopted strategy is to use vehicle buffers to absorb the fluctuation. When buffers are full, there are only two ways that can be taken to minimize the inventory cost; one is to lower down retail price to stimulate demand, the other is to stop production to reduce stock, either one implies high cost.
- Build on order is the best solution, but seasonality is a major issue. Manufacturing division is unable to meet some seasonal demands, unless huge overtime is given.
- Right material, and right quantity at right time; the challenge is the long lead time required, including all process time. Some materials required longer than 6 months' lead time. In the mean time, entire market may have severe fluctuation to maintain low inventory and meet the marketing demand is a big challenge.

**Materials Planning and Logistics, Ford India, Chengelpet, Tamil Nadu, India.*

- Handle engineering change precisely, without incurring any obsolescence cost. There are various reasons for engineering changes taking place, some of them are marketing activities, cost down ideas, process improvements, quality issues, etc. Most of the time, engineering changes are implemented on very urgent basis. How to minimize scrap and meet all the above requirements. This needs huge efforts.
- Movement of material from different sources into the main plant, and transporting built-up vehicles to dealers involves big cost. To reduce the cost is a major challenge. Also, in particular, in future, global sourcing is going to be a trend. How to find lowest cost and shortest transit time route is also a challenge.
- Record integrity is one more challenge; there are so many links to keep good record. Ensure book quantity is equal to actual quantity, such as receiving key-in, usage, scrap handling, etc. If there is any broken link, record integrity will be a big trouble.

FUTURE VISION FOR AUTOMOTIVE MANUFACTURING COMPANIES

1. With the continuous improvement in manufacturing technologies, all the materials must have lead time as in the electronic industry, then customer orders can be fulfilled within short time. With this achievement, all material planning and logistics personnel can get rid of the nightmare of having too much of inventory on hand.
2. Industrywise joint logistics planning—all car makers should compete on brand and service in terms of logistics; all car makers should work together to standardize processes, jointly purchase materials from common source, to have higher leverage for cost reduction.
3. More and more standard parts should be used industrywise to reduce purchasing cost and engineering change risks. If this works out, all suppliers can set up warehouses near the car manufacturing plant.
4. Assistance to improve supplier capability: Material planning staff should not focus on chasing materials only. Every one in this field should be experienced in all manufacturing areas, logistics, and material handling so that they can assist suppliers on process and productivity improvement and layout, rack design to reduce transportation cost, and unit price. To achieve this objective, all the material planning and logistics staff should be prepared for job rotation, to understand the nature and operation of different functions. The head of any material planning and logistics section should be responsible for imparting training to the people and rotate them to ensure that everyone is strong enough to independently assist suppliers.
5. Supplier communication is a very important link for getting right material at the right time. Internet should be used for all communication, even for video communication to reduce communication cost drastically, such as order release and stock on hand. Production scheduling should be exposed to suppliers through internet to ensure better partnership between car maker and suppliers.
6. With the advancement of technology, all parts should carry a transmitter (RFID Tag) and automatic booking is the final solution for maintaining effective record integrity.

17

Impact of Globalization on Products and Manufacturing Engineering Education and Practices

*Dr. M.P. Chowdiah**

INTRODUCTION

The impact of globalization on products and manufacturing education and practice in the present liberalised economy of the world has resulted in an integration of product, process, design, quality assurance, and equivalent stages of manufacturing education and practice and policy of product design. The concept, strategy and policy for the impact of globalization and different elements involved in product design and manufacture for globalization are discussed here.

In the globalization of economy lifting trade barriers, converting local trade into global market, blending of engineering and technical knowledge with information at the global level have all contributed to the zenith of quality assurance in all the manufacturing activities, such that the product is reliable and suitable for all types of working conditions, remain durable with long period of service utilising least power, energy and cost. This requires sound knowledge of product design acquired by engineers with the desired education and practice at all levels of manufacturing system, which leads to globalization in the liberalised economy.

MANUFACTURING TECHNOLOGY AND ENGINEERING

Technology and engineering as applied science encompasses several divisions, such as agricultural engineering, architectural engineering, aerospace engineering, etc. as shown in Fig. 17.1 that contribute to manufacturing directly or indirectly.

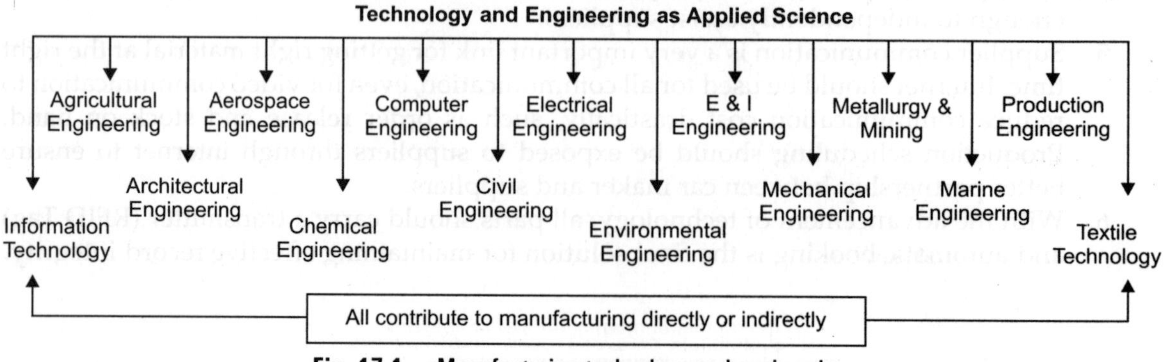

Fig. 17.1. Manufacturing technology and engineering

**Former President, The Institution of Engineers (India), Kolkata, India.*

The impact of globalization on manufacturing leads to global competitiveness which, in turn, leads to cost effective quality customised products globally and thus to agile manufacturing as shown in Fig. 17.2.

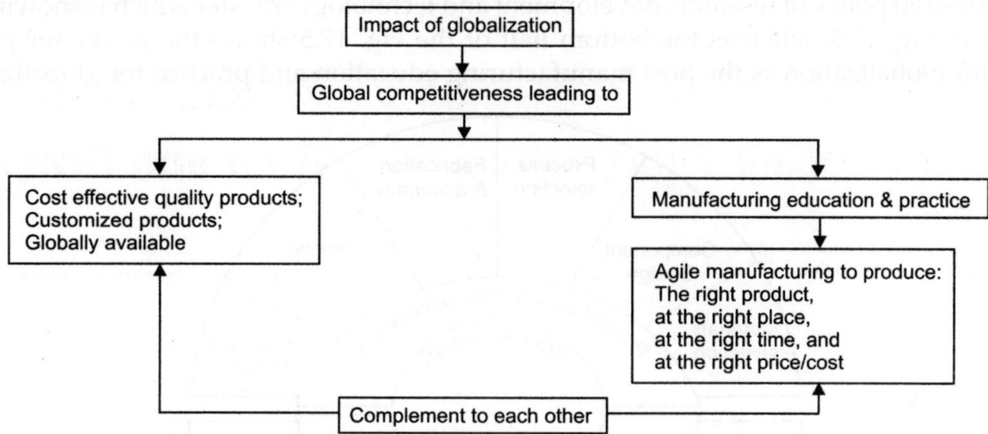

Fig. 17.2. Impact of globalization on manufacturing and practice

IMPACT OF GLOBALIZATION ON PRODUCT DESIGN

The globalization impact on product design is an integration of product design policy with equivalent stages of manufacturing, education and practice for customised product with quality assurance as shown in Fig. 17.3.

The product quality assurance design demands the quality at all stages of design, materials selection, established processes and fabrication techniques, assembly, testing, packing and finally marketing as indicated in the upper portion of the Fig. 17.4 whereas the lower part of the Fig. 17.4 shows procedure to execute the standard product design in which the workforce must be injected with quality education based on source of control such as admission criteria, curriculum design, programme selection, curriculum implementation, evaluation and employability, etc.

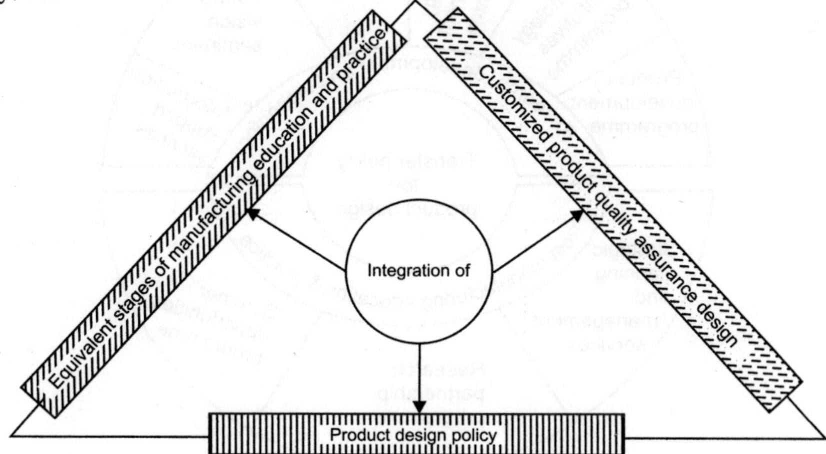

Fig. 17.3. Globalization impact on product design

In view of economic reforms based on policy of globalization which affects the industrial and economic scene by increasing competition in quality and practice of the products and services, they must prove to be of highest quality and yet competitive. It is therefore necessary to have desired policy of research, development and technology transfer which is shown on the top half of Fig. 17.5, whereas the bottom half of the Fig. 17.5 shows the policy for product design for globalization as the post manufacturing education and practice for globalization.

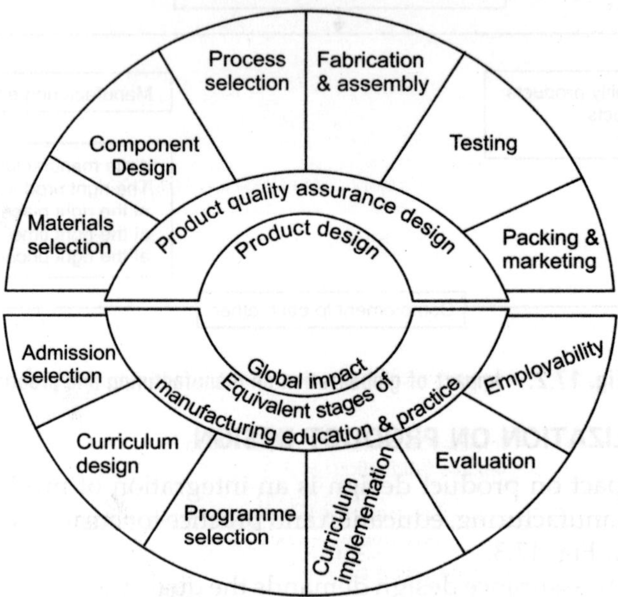

Fig. 17.4. Globalization impact on product and manufacturing education and practice

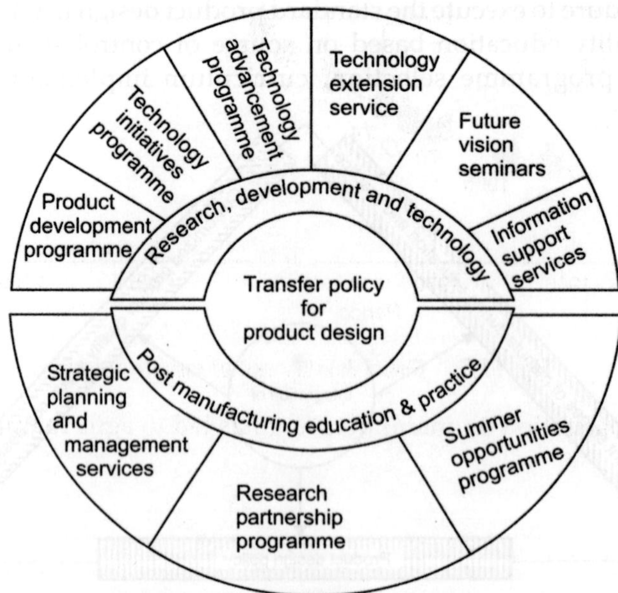

Fig. 17.5. Globalization impact on desired policy for product design and manufacturing

IMPACT OF GLOBALIZATION ON MANUFACTURING AND PRACTICE

Before the discussion of the changes in manufacturing for globalization it is necessary to discuss the global development of manufacturing from 1940 to 2020 which is shown in Fig. 17.6. Further, the sequential scenario of manufacturing in the world presents the history of manufacturing from the past to the present status as shown in Fig. 17.7.

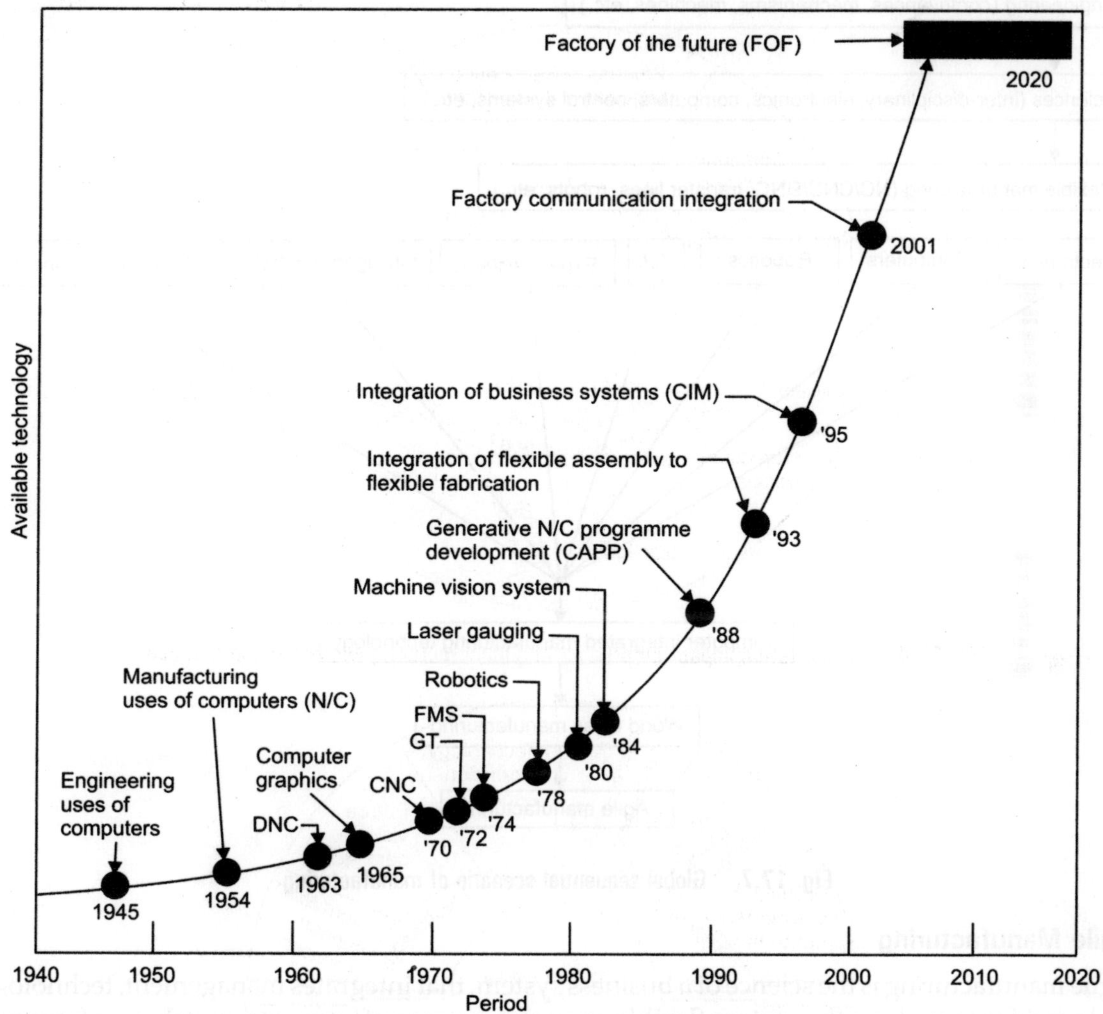

Fig. 17.6. Global development of manufacturing

The impact of globalization on manufacturing has led to agile manufacturing, which may be defined as follows:

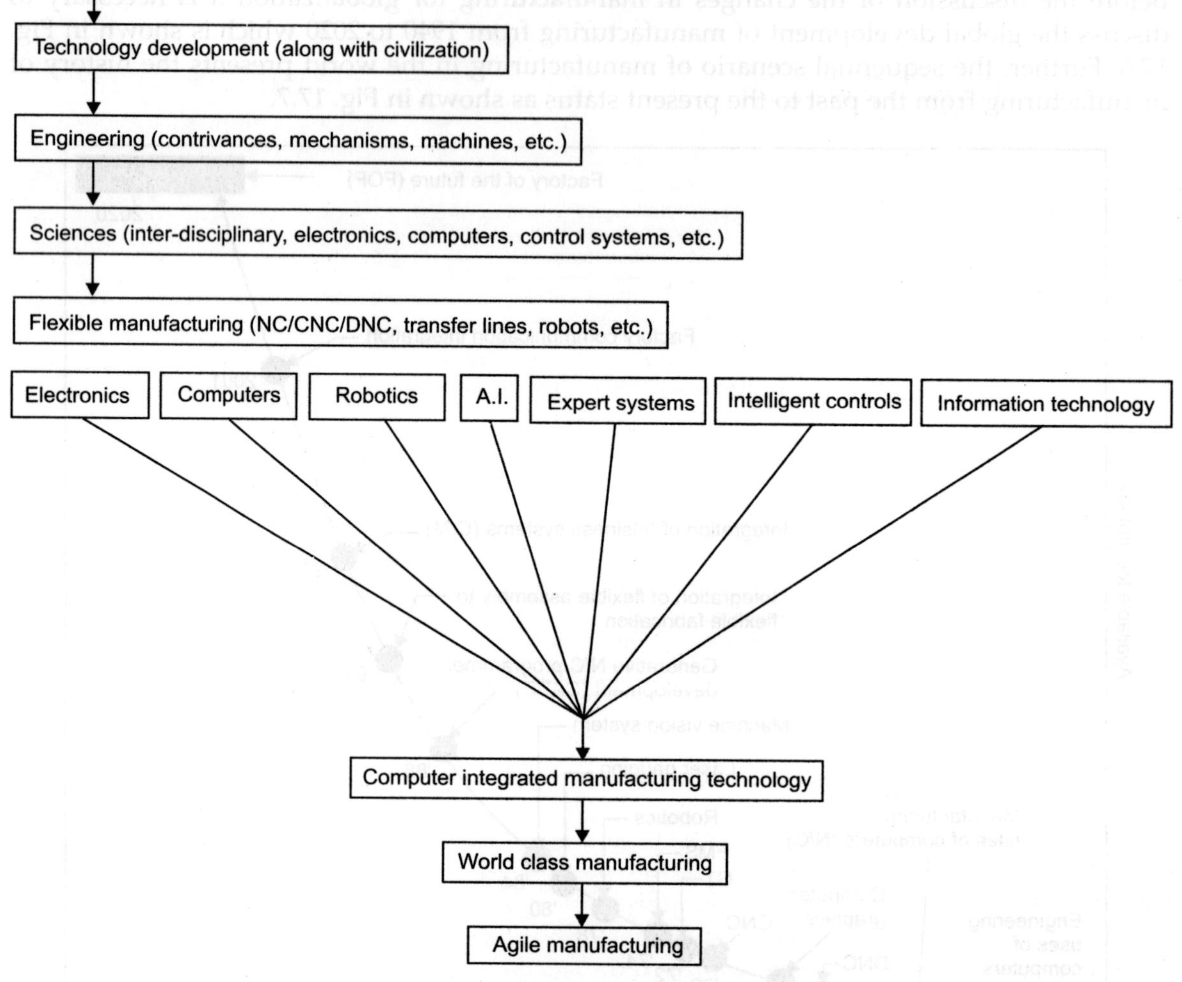

Fig. 17.7. Global sequential scenario of manufacturing

Agile Manufacturing

Agile manufacturing is the science of a business system, that integrates management, technology and workforce, making the system flexible enough for a manufacturer to switch over from one component/product that is being produced to another component/product that is desired to be manufactured in a cost-effective manner, in a short-time, within the framework of the system.

Agile Manufacturing Approach

The goal of agile manufacturing approach (AMA) is to discover and codify guiding principles for manufacturing and educating future leaders for manufacturing workplace and otherwise infuse important principles and technologies into manufacturing practice.

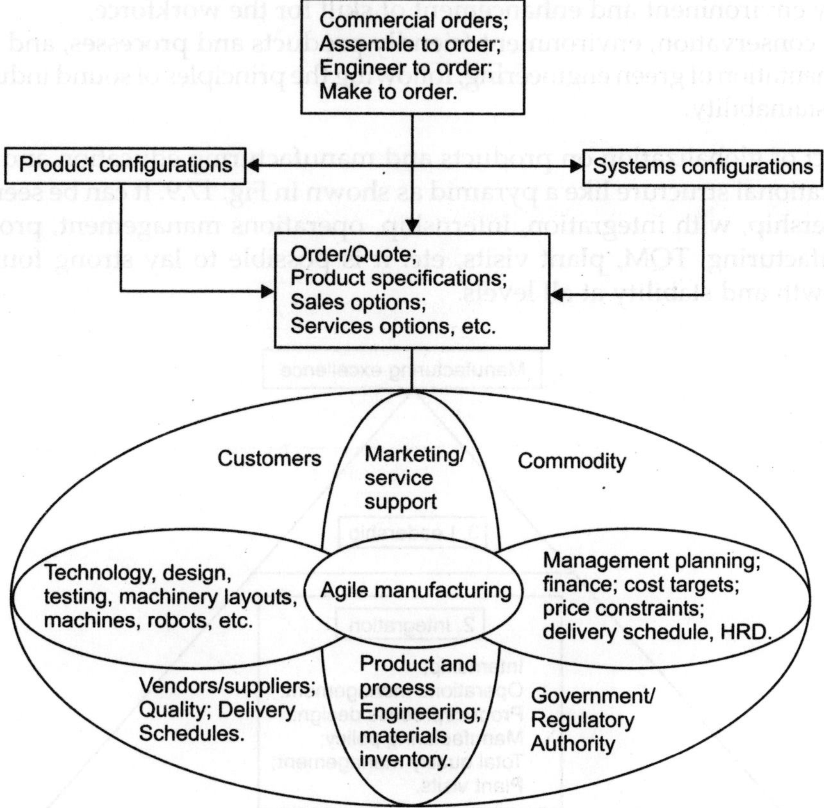

Fig. 17.8. Agile manufacturing due to globalization

The manufacturing has to be viewed as a broad-based activity ranging from product design through manufacturing, product use, maintenance, etc. But the AMA (Agile Manufacturing Approach) has been recognised as an interdisciplinary activity that requires the seamless integration of technology, management, information system, workforce, etc. Agile manufacturing has got an holistic approach rather than taking rifle-shots at individual issues, to achieve quick fixes by bridging the traditional technology, management, workforce, etc. with a broad understanding of manufacturing that integrates key functions and disciplines involved in creating, designing, making, selling/servicing products, etc. It encompasses not only the critical operations like technology, product and process engineering, administration and marketing/sales/services within a corporation but also vendors/suppliers, customers, and community at large and government as shown in Fig. 17.8.

Meeting Globalization Impact

In order to meet globalization impact, the following factors need to be considered:

- Supply of customer oriented products,
- Challenges posed by multinational companies, which are already practising the concept of agile manufacturing,
- Need to produce components of international standards, in terms of quality and cost,

- Healthy environment and enhancement of skill for the workforce,
- Energy conservation, environment friendly products and processes, and
- Implementation of green engineering, following the principles of sound industrial ecology and sustainability.

The impact of globalization on products and manufacturing education and practices has led to an educational structure like a pyramid as shown in Fig. 17.9. It can be seen that with an effective leadership, with integration, internship, operations management, product/process design, manufacturing, TQM, plant visits, etc. it is possible to lay strong foundation for a sustained growth and stability at all levels.

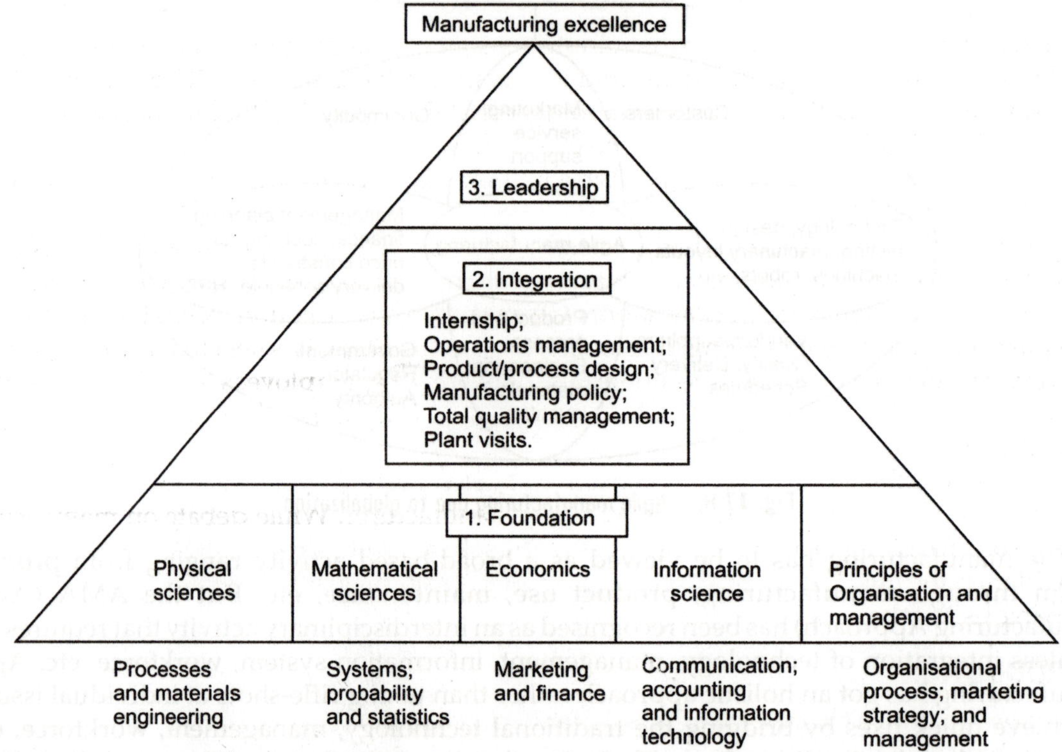

Fig. 17.9. Impact of globalization on manufacturing education and practice

18

Information Technology as a Change Enabler

*Dr. M.P. Ravindra**

The pace of life is speeding up by the minute—a sign of changing times and habits of people world over. This is possibly inevitable and irreversible. Where all this will lead us, it is not for us to say. But, it is a truth that stares us in the face that those people who cannot keep pace with the change and adapt to the same will be completely forgotten.

The only difference is that it won't take a million years. It will be in the first ten years of the new Millennium.

The information revolution that is pushing everything around us into a roller coaster has caught manufacturing technology too. Nearly every item of merchandise will be moving closer to *made to order or engineer to order* situation. Given the way information technology is changing every aspect of modern day living, it is foolish to fight the onset. It is best that the management of every conceivable business is to undertake educating the employees at all levels about the inevitability, so that they willingly join forces to harness IT to transform their business into a vibrant enterprise.

This chapter brings out some perspectives of this powerful change enabler and need for embracing it for survival and growth by every manufacturer. While debate on many aspects may go on, accepting IT will help us focus on the opportunities round the corner. Else, again we will miss the bus. India has a great opportunity, as we have not invested much on IT so far allowing us to take a quantum jump.

INTRODUCTION

It has now become a cliché—'change is the only permanent thing'. But at the cost of repetition we find how true this is. It is only how people have adopted and adapted to change that has been decisive in determining how different societies have performed or perished.

This process of adaptation has been forced by nature in most cases due to physical and natural boundaries. But man has fought it ingeniously by acquisition of information, converting it into knowledge by wise persons using tools available at their disposal. This called for ability to invent tools and develop skills to use the same, no matter whether you created the tool or not.

It is no different in the new millennium. The ensuing millennium is dubbed to be the information era where knowledge will be a power and a resource. It is therefore imperative that we quickly master the various information technology tools currently and those that will be available in future and exploit it to the fullest to transform the businesses we are running.

*General Manager, Ramco Systems, Chennai, India.

It is thus evident that information technology will play a pivotal role in this transformation and will enable the change agents in enterprises to bring about necessary changes quickly and with least disruption to the users of the services of one's organisation.

We will discuss how possibly one can go about exploiting the advances in IT for business success of the enterprise and benefit society at large.

BUSINESS AS ART OF TRANSFORMATION

From a thirty thousand feet perspective of the society, one can see orderliness in all the chaos present. By the very nature of human beings, a person cannot be peaceful with routines. He seeks variety in food, sports, music, work and tools used practically in everything he does. Therefore, human beings demand that new products and services are available to him in an easily 'accessible' manner, whether he uses it or not. He wants 'choice'.

Therefore, it is essential for the 'provider (read manufacturer)' of the products or services to innovate and bring out new and attractive choices faster and cheaper and place it in front of the customer in the fastest way.

It is easy to see that the people who are in any business are *actually transforming raw inputs/ideas into products or services*. Changes have come in inputs, tools and skills used and also the manner in which the end product is delivered and charged.

If one looks at the evolution of mankind, one will observe that from anthropological times *using tools* has been the key differentiation between man and animals. Only those who ran fast or used weapons that could be deployed from a distance and travelling fast hunted successfully. Ability to sniff and change directions or the ability to read the 'weather' quickly could be successful in surviving. Else they became the hunted, those who could not be perished; not even becoming a part of history.

Have things changed now, let us ask ourselves? Nothing has changed. Only those who recognise this, will prepare themselves for the coming onslaught of those who are arming themselves with the new weaponry will survive and it is only paranoids who will survive. The new tool in today's global economy is information technology.

It is established that only those who will master information technology tools in all its facets can compete in the new world order that is getting established.

Now, let us come to the main theme—IT as a change enabler. We will also discuss the areas of manufacturing technology which has touched and is likely to alter in future.

PERVASIVENESS OF IT IN MANUFACTURING

We can divide the manufacturing business into design, production, logistics, sales and delivery. These can be reduced to two categories, namely, commercial aspects and manufacturing/production aspects:

1. The application of IT in the commercial areas is too obvious with the proliferation of Enterprise Resource Planning (ERP) products/softwares as also point products for specific areas, named variedly as NW, MMS, PM, etc. However, the application of IT in production areas may be detailed. Some of the business benefits that accrue by use of IT are also listed below. All of them are centered on customer satisfaction.
   (a) First, the engineering design today is totally transformed. The number and variety of specialized tools for CAD are many—from simple to very sophisticated. Design

centres can be spread out or even out-sourced to reduce costs. This will also give access to best of the design centres from across the world enabling reduction of 'time to market'.

(b) Also, non-destructive testing (NDT) methods using IT tools have become possible.

(c) Accepted designs can be made available to production units over internet, or intranet and to low cost production facilities globally.

(d) Manufacturing facilities can be located closest to the most critical inputs independent of the location of R&D facilities (Dell, IBM are live examples).

2. In the shop-floor multiple levels of IT, usage from point of view of direct business benefits is possible:

   (a) The use of bar code to DCS (Distributed Computer Systems) to control the machine operations and for quality checking is available today. This has resulted in reducing the number of rejections, saving costs in material and rework.
   (b) Further, flexible manufacturing systems (FMS) have helped in meeting customer demand both for Deliver to Promise and special requirements by accepting changes to order till last minute.
   (c) Servicing small lot orders.
   (d) Reduction of time spent by shop-floor personnel on non-conformity reporting and instead focuses on improving quality and productivity.
   (e) Ability to take proactive remedial measures to increase customer loyalty by implementing systems for lot tracking.

3. These apart, IT tools can help in the following areas.

   (a) Increase market reach using internet, intranet, and extranet (extended supply chain complementing ERPs)
   (b) Optimise the cost of operations
   (c) Cost of material by reducing inventory
   (d) Minimizing down time of captive resources
   (e) Reducing cost of maintenance
   (f) Use the right teams for tasks
   (g) Reduce rejections
   (h) Reduce working capital costs
   (i) Benchmarking
   (j) Proud and happy workforce.

The list above is illustrative. Innovative organisations would go about creating and upgrading a good IT infrastructure to get a control on the metrics and make continuous adjustments to stay ahead of competition. Manufacturing industries are especially vulnerable to changes in market requirements and product life cycles are getting reduced and will call for quick responses.

The only way is to acquire and master the information technology tools in the data warehousing and data mining. This will help in tracking even slight shifts in customer behaviour and adjust the marketing and production strategy accordingly.

Independent of the tools used for change, it is important to remember, as Bill Clinton said during the Palestine National Congress 1998 that "It takes time for people to change. It takes even more time for people to get benefits of the change."

It is therefore critical to get support from the people for the change through well thought out change management processes and training.

There is an incident from Albert Einstein's classroom during an examination:

Student asked Albert Einstein: "Sir have you noticed that the questions are the same as last year?"

Albert Einstein told the students: "Yes! But the answers are all different this year."

19

Development of Robotic Systems for Use in the Hazardous and Constrained Areas

*Dr. M.S. Ramakumar**

The use of robots in the manufacturing industry is a matter of reality. There are still certain applications where the skill and flexibility of a person cannot be replaced. But in environments, where access is difficult or forbidden owing to area constraints or hazards, the use of robots can be the only answer.

The presence of radioactivity (in a nuclear facility), explosives, work in outer space or deep under-sea pose difficult environments and prevent long term presence of human beings to carry out detailed work. Presence of dimensional constraints prevents direct access to perform tasks.

Robots and manipulators have to be specially developed to perform such a whole variety of tasks. Systems will have to be designed, developed, tested and utilised for specific requirements. Transparency in the system is important when robot operate in remote environments. There must be adequate feedback to the operator to enable him to exercise proper control over the operation. Kinesthetic feedback, real-time simulation and virtual reality can contribute not only to robot operation but also to design and development.

Keeping the above points in mind, it is necessary to modularise and standardize components and subsystems. This will enable rapid assembly of needed configurations, help in quick maintenance and avoid problems arising out of obsolescence.

The designers of facilities where robots are deployed should interact closely with robotic system developers to evolve facility and plant designs which provide certain necessary features which will enable rapid and efficient development of robotic configurations to perform operations where one cannot have access.

**Director, Automation and Manufacturing Group, Bhabha Atomic Research Centre, Mumbai, India.*

20

A Hybrid Method to Find Fracture Parameters K_I and K_{II} to Assist Punching Operation Using FEM and Photoelasticity Techniques

G. Ganesan[1], K.R.Y. Simha[2] and M. Adithan[3]

A new hybrid method is developed to aid in the determination of mixed mode stress intensity factors via the photoelastic technique in conjunction with finite element method. Making use of stress optic law in association with FEM, the hybrid method is carried out using an edge notched specimen. This hybrid method is used to calculate K_I, K_{II} and σ_{ox} for different crack lengths and crack orientations.

Nomenclature

a	Crack length
h	Thickness of the Model
S.I.F	Stress Intensity Factor [MPa $\sqrt{m}$]
K_I	Stress Intensity Factor in mode-I
K_{II}	Stress Intensity Factor in mode-II
σ_{ox}	Far Field Stress (MPa)
σ_x, σ_y, τ_{xy}	Stresses Parallel to X, Y Coordinates
τ_m	Maximum Shear Stress
N	Fringe Order
f_σ	Material Fringe Value [MPa-m/fringe]
r, θ	Polar Coordinates
FEM	Finite Element Method
LEFM	Linear Elastic Fracture Mechanics

[1]*Assistant Professor, Mechanical Engineering, Vellore Institute of Technology, Vellore, India.*

[2]*Associate Professor, Mechanical Engineering, Indian Institute of Science, Bangalore, India.*

[3]*Professor, Mechanical Engineering, Vellore Institute of Technology, Vellore, India.*

INTRODUCTION

The stress field around crack tips in punching operations are complex due to contact stress effects. For such a problem, finite element method can provide additional information to extract the SIFs from the state of stress around the crack tip for various clearances between the punch and die. Though a number of methods are used to calculate K_I and K_{II} by FEM, a hybrid method explained here is simpler in many ways compared to existing FEM methods. Calculation of K_I and K_{II} along with σ_{ox} for two-dimensional fracture problems is a significant feature in this method. The results thus can be compared with the photoelastic results. This hybrid method enables computer simulation of isochromatic fringe patterns observed in the photoelastic investigation. FEM results are mainly employed to establish the stress field in the crack tip vicinity. Stress optic law is then invoked to plot the isochromatics and extract the mixed mode fracture mechanics parameters.

CONCEPT OF THE NEW METHOD

FEM is widely used in many areas of research. In fracture mechanics its use is growing rapidly. In FEM, model of the problem is divided into number of finite elements. Governing equation of the elements is given by the stiffness matrix, displacement vector and load vector. Suitable elimination method is used to solve simultaneous equations. From the displacements of the nodal points stresses are calculated. Using the FEM technique principal stresses at the Gauss point locations of the element are calculated. Thus, coordinates and stresses of the Gauss points are obtained. Cartesian coordinates of the Gauss points are transformed to local polar coordinates referred to the crack tips. The origin is also shifted to the crack tip.

The present hybrid FEM method is based on displaying, from FEM results, maximum shear stress contours. These contours represent isochromatics in an equivalent photoelastic model. Consequently, it is possible to exploit the overdeterministic method to extract K_I and K_{II}. Additionally, as a bonus, the new hybrid FEM method also yields the value of σ_{ox}. In the zone of interest limited to certain area of the crack tip, recalling the photoelastic equation governing the mixed mode fracture.

$$\left(\frac{N_k f_\sigma}{h}\right)^2 = \frac{1}{2\pi r_k}\left[(K_I \sin\theta_k + 2K_{II}\cos\theta_k)^2 + (K_{II}\sin\theta_k)^2\right] +$$

$$\frac{2\sigma_{ox}}{\sqrt{2\pi r_k}}\sin\frac{\theta_k}{2}\left[K_I \sin\theta_k(1+2\cos\theta_k) + \left(1+2\cos^2\frac{\theta_k}{2}+\cos\theta_k\right)\right] + \sigma_{ox}^2 \quad (1)$$

$K = 1, 2, 3, \ldots$

The validity of the equation is limited to the close vicinity of the crack. The crack tip zone where the equation is valid is given by a fraction of the crack length. Near the crack tip the stress field is not affected much by the contact stress. However, care has to be exercised to see that the information from the fringes are obtained only for those fringes which are not disturbed by the contact stress field. From the fact that the fringes are the result of suppression of light rays along the maximum shear stress plane ($(\sigma_1 - \sigma_2)/2 = \tau_m$), the following equation obtained from stress optic law is given below:

$$(2\tau_m)^2 = (\sigma_x - \sigma_y)^2 + (2\tau_{xu})^2 = \left(\frac{Nf_\sigma}{h}\right)^2 \quad (2)$$

From the principal stresses at the Gauss points the fringe order N can be calculated. The polar coordinates and the fringe order for 20 or 30 Gauss points around the crack tip are used in the overdeterministic method. The iterative overdeterministic method gives the corrected values of K_I and K_{II} and σ_{ox} so that the equation 1 is satisfied.

BENCHMARK PROBLEM

To demonstrate the new hybrid method, a single edge notched specimen, whose mesh generation and boundary conditions, as shown in Fig. 20.1, under tension, is carried out. The model under

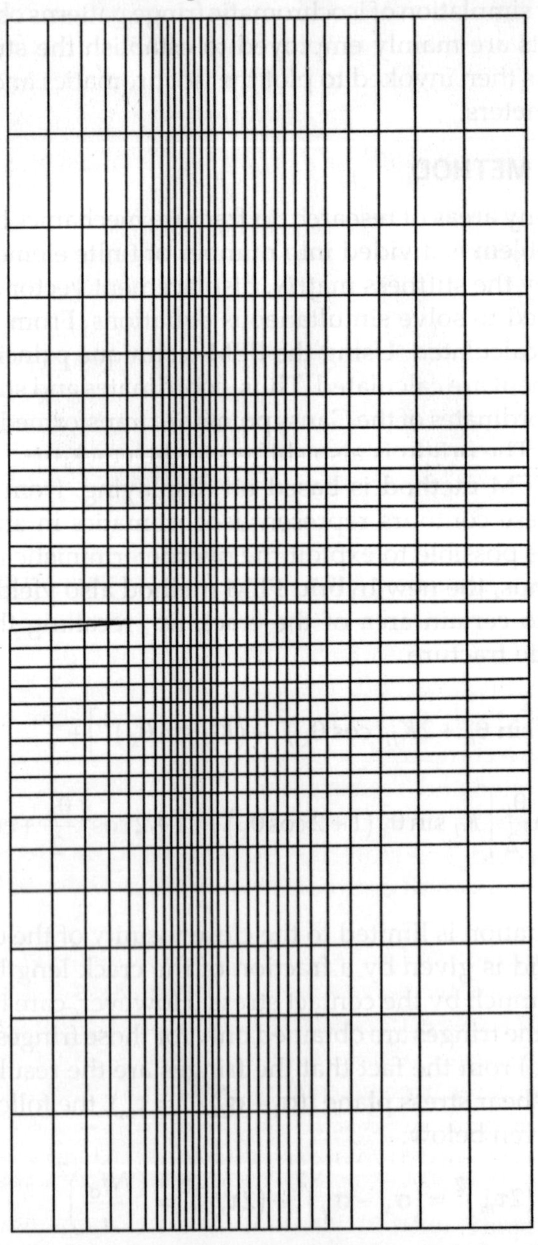

Fig. 20.1. Numerical model of the single edge notched specimen

Fig. 20.2. Isochromatic fringe pattern (Single edge notched specimen)

tension is photographed to get isochromatic fringe patterns (Fig. 20.2). Analysis of the model is treated using the new hybrid method. The values of K_I and K_{II} and σ_{ox} for the model using the overdeterministic method and the new hybrid method are used to reproduce the isochromatic fringe patterns and they are compared.

Crack length, load applied and Young's Modulus are the data used in the FEM to find the maximum shear stress contours. From the shear stress contours polar coordinates r, θ of about 30 points are carefully selected. The material fringe value 'f_σ' of the photoelastic specimen and thickness 'h' are the other important data to be furnished in order to find K_I, K_{II} and σ_{ox}, of the model under test.

The deformed configuration of the model of the benchmark problem is shown in Fig. 20.3 and the isochromatic fringe pattern reproduced from the hybrid method and experiment are shown in Figs. 20.4 and 20.5 respectively. In the benchmark problem conducted to verify the validity of the new hybrid method shows that the result is matching well with the numerical method based on the overdeterministic method. The value of K_I obtained from the Hybrid method is 0.739 MPa$\sqrt{m}$ and the value of K_{II} obtained from the experiment using overdeterministic method is 0.646 MPa$\sqrt{m}$. The difference in the value of K_I between these two methods, though about 10 per cent, cannot be concluded as the error in the new hybrid method because the photoelastic model used to compare the result may induce some error due to the

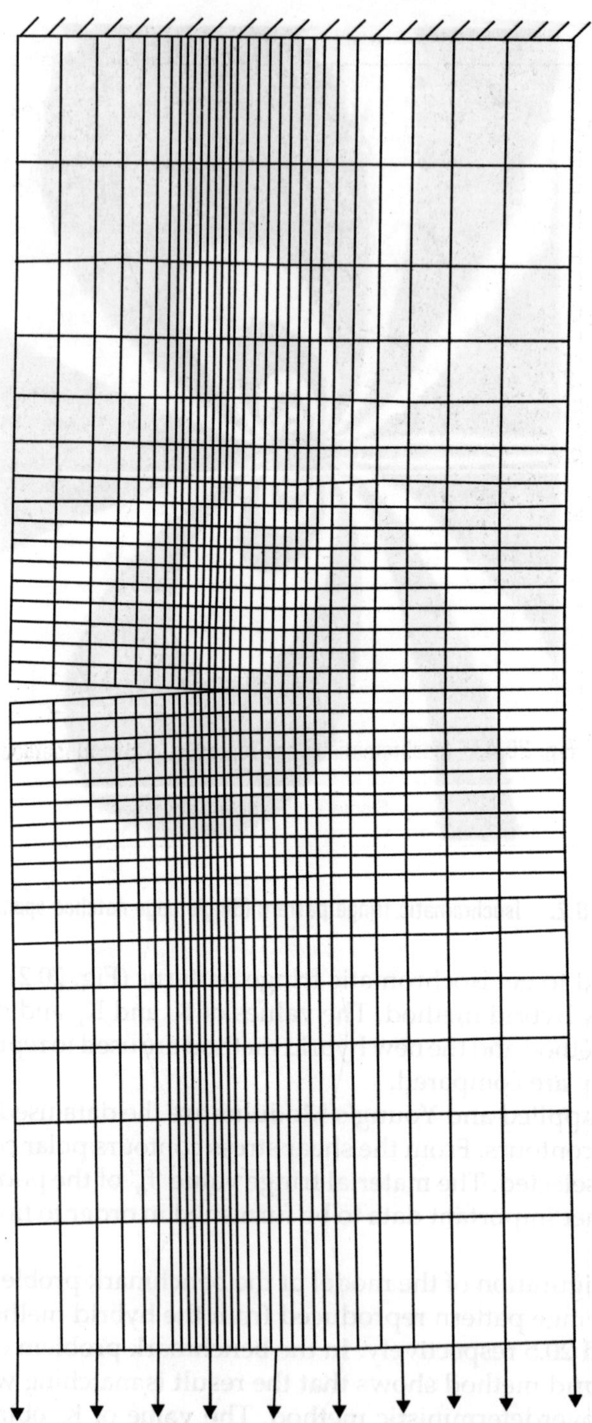

Fig. 20.3. Deformed configuration of the numerical model

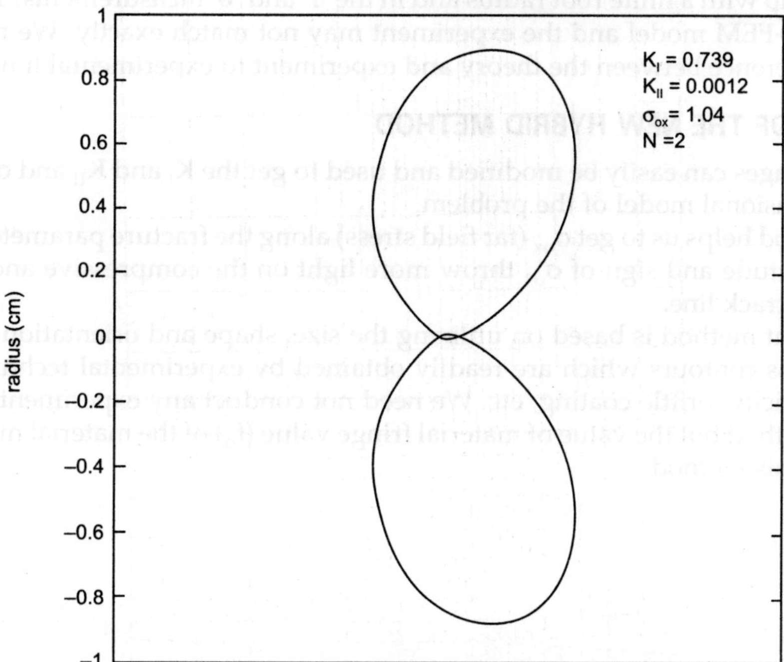

Fig. 20.4. Isochromatic fringe reproduced (Hybrid method)

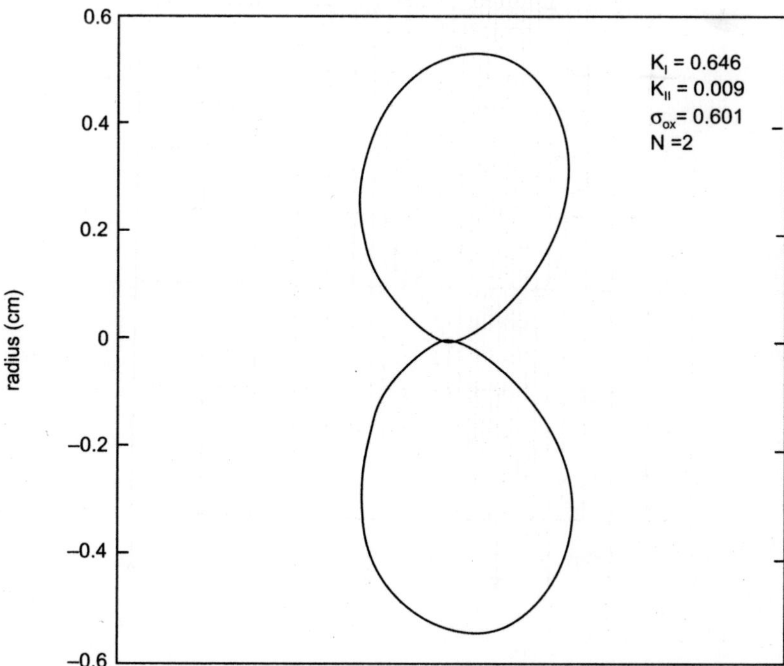

Fig. 20.5. Isochromatic fringe reproduced (Experimental)

non-ideal crack tip with a finite root radius and in the 'r' and 'θ' measurements. The boundary conditions of the FEM model and the experiment may not match exactly. We may therefore attribute the difference between the theory and experiment to experimental limitations.

ADVANTAGES OF THE NEW HYBRID METHOD

1. FEM packages can easily be modified and used to get the K_I and K_{II} and σ_{ox} values of a two-dimensional model of the problem.
2. This method helps us to get σ_{ox} (far field stress) along the fracture parameters K_I and K_{II}. The magnitude and sign of σ_{ox} throw more light on the compressive and tensile field along the crack line.
3. The present method is based on utilizing the size, shape and orientation of maximum shear stress contours which are readily obtained by experimental techniques such as photoelasticity, brittle coating, etc. We need not conduct any experiment to apply this hybrid method but the value of material fringe value (f_σ) of the material must be known to apply this method.

21

Software Reliability Estimation Using Exponential and Gamma Models

Rashpal S. Ahluwalia* and Mitesh Parekh*

Here we describe the use of the Exponential and the Gamma models to estimate software reliability. The models were applied to sample software failure data. The model parameters were estimated using the maximum likelihood estimation technique. For the sample data set the Gamma model provides better estimation of software reliability because of its flexibility to handle increasing and decreasing failure intensity.

NOTATION

τ	Execution Time (CPU time)	t	Calendar Time
TET	Time at End of Test	TTF	Time To Failure
TNF	Total Number of Failures at End of Test	b0	Estimated Faults in the Software
DFI	Desired Failure Intensity	b1	Estimate of Failure Decay Factor
TDFI	Total to reach DFI	CFI	Current Failure Intensity
CMV	Current Mean Value	DMV	Desired Mean Value
$\mu(\tau)$	Mean Value Function	$\lambda(\tau)$	Failure Intensity Function

INTRODUCTION

The size and complexity of computer-intensive systems has grown dramatically, and the trend is likely to continue certainly in the future. Some of the examples of highly complex hardware and software systems can be found in the telecommunications industry where call switching is supported by hundreds of millions of lines of source code. In the avionics industry, almost all-new payload instruments contain their own microprocessor system with extensive embedded software. The operating systems on today's office and home computers consists of 1 to 5 million lines of source code along with many other shrink-wrapped application software packages of similar size. The demand for complex hardware and software systems has increased more rapidly than the ability to design, implement, test and maintain them. As dependency on computer increases, the possibility of crises from a computer failure also increases. The impact of such failures can range from inconvenience (e.g., malfunctions of home appliances), economic damage (e.g., interruptions of banking systems), to loss of life (e.g., failures of flight systems or

**Industrial and Management Systems Engineering, College of Engineering and Mineral Resources, West Virginia University, WV, USA.*

medical software). Therefore, software reliability needs to be specified and estimated, at least for safety critical systems.

Software Reliability is defined as the probability of failure free software operation during a specified period of time in a specified environment. It is one of the key attributes of software quality, a multidimensional property including other factors like functionality, usability, performance, serviceability, capability, instability, maintainability and documentation. Reliability is also an essential ingredient in customer satisfaction for commercial, as well as governmental organizations. The ISO 9000–3 guidelines specify measurement of field failures as the only required quality metric: "*. . . At a minimum, some metrics should be used which represent reported field failures and/or defects from customer's point of view. . . . The supplier of software products should collect and act on quantitative measures of quality of these software products.*"

The quality of the software needs to be quantitatively defined from customers point of view by defining failures and failure severity and by determining a reliability objective and specifying balance among key quality objectives, such as reliability, delivery, and cost. The customer usage can be quantified by developing an operational profile. Operational profile is a set of disjoint alternatives of system operation and their associated probabilities of occurrence. The construction of operational profile encourages testers to select test cases according to the systems operational usage, which contributes to more accurate estimation of software reliability in the field.

Moreover, reliability during testing is done to meet the required reliability of the software. Typically, a reliability model is built by assuming suitable probability density functions for the random variables of interests. Using the reliability models, the best estimates are obtained. These estimates are then used to project the reliability during filed operations in order to determine if specified reliability has been met. This procedure is an iterative process since more testing will be needed if the objective is not met. A software reliability model specifies the general form of the dependence of the failure process on the principal factors that affect it: fault introduction, fault removal, and the operational environment. Fig. 21.1 shows the failure intensity and mean value functions, the building blocks of software reliability modeling. The failure rate of a software system is generally decreasing due to the discovery and removal of software faults. At any particular time (say, the point marked "present time"), it is possible to observe the history of the failure rate. Software reliability modeling forecasts the curve of the failure rate by statistical evidences.

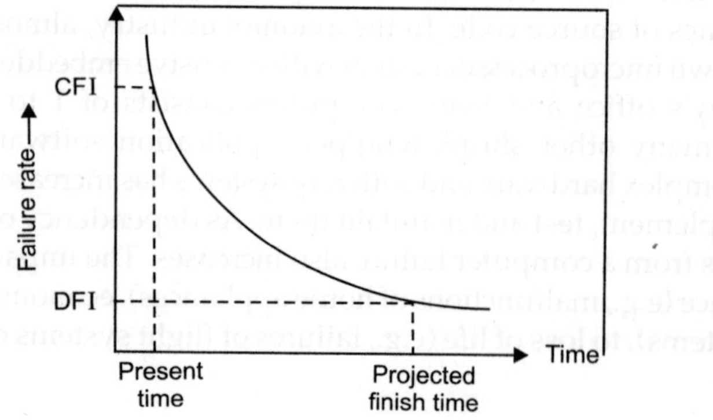

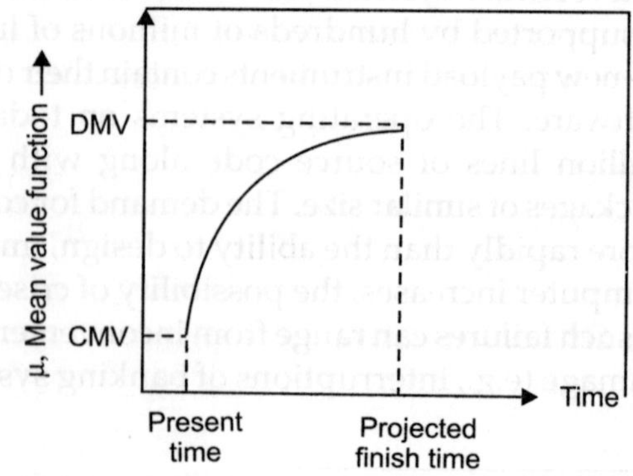

Fig. 21.1. Failure intensity and mean value functions

Reliability quantities have usually been defined with respect to time, although it is possible to define them with respect to other variables. Time may be considered as (a) Execution Time, (b) Calendar Time, or (c) Clock Time, i.e. the sum of times from program start to program end, without counting shut down periods. As shown in Fig. 21.1, the two functions can be defined for the time variation of a random process. The *mean value function* (μ), as the average cumulative failures associated with each time point; and the *failure intensity function* (λ) as the rate of change of mean value function or the number of failures per unit time. Note that the failure intensity function is the derivative of the mean value function. When there are no changes being made in the software product, i.e. no debugging or software corrections, then λ is constant and failures occur according to a *homogeneous random process*. During software debugging, a *non-homogeneous process* (NHP) takes place. Figure 21.1 shows a typical non-homogeneous process in the sense that the mean number of failures experienced increases with time in such a manner that the failure intensity decreases.

THE EXPONENTIAL MODEL (Basic Musa Model)

The Basic Musa model is used widely for software reliability modeling. It was developed at the AT and T Bell Laboratories by John Musa. Musa was one of the first to use the actual execution time of the software component on a computer for the modeling process. The assumptions and the data requirements for the model are:

1. The cumulative number of failures by time τ, $\mu(\tau)$, follows a Poisson process with mean value function $\mu(\tau) = b0^*(1 - e^{-bl\tau})$, where, b0, bl > 0. The mean value function is such that the expected number of failure occurrences for any time period is proportional to the expected number of undetected faults at that time. Since $\lim \tau \to \infty\ \mu(\tau) = \lim \tau \to \infty\ b0^*(1 - e^{-bl\tau})$ it is a finite failure model. The parameter b0 is the total number of faults that would be detected in that limit.
2. The execution times between the failures are piecewise exponentially distributed, i.e. the hazard rate for a single fault is constant. This is why the model belongs to the exponential class.
3. The quantities of the resources (number of fault-identification, correction personnel and computer times) that are available are constant over a segment for which the software is observed.
4. Fault-identification personnel are fully utilized and computer utilization is constant.
5. Fault-correction personnel utilization is established by the limitation of fault queue length for any fault-correction person. Fault queue is determined by assuming that fault correction is a Poisson process and that servers are randomly assigned in time.

Assumptions 3 through 5 are needed only if linking of execution time to calendar time is desired. The data requirement to implement the model is that the actual times at which the software failed, $\tau_1, \tau_2, \ldots \tau_n$ or the elapsed time between failures $x_1, x_2, \ldots x_n$, where $x_i = \tau_i - \tau_{i-1}$ be known.

THE GAMMA MODEL (S-Shaped)

In the S-shaped reliability growth model the per-fault failure distribution is according to the gamma distribution. The number of failures per time period, however, is a Poisson process. It is a finite failure model, i.e. $\lim \tau \to \infty < \infty$. It is patterned, as the mean value function is often a characteristic S-shaped. The basic assumptions and the data requirement for the model are as follows:

1. The cumulative number of failures by time τ, $\mu(\tau)$, follows a Poisson process with mean value function $\mu(\tau) = b0*(l - (1 + bl*\tau) *e^{-bl\tau})$ for b0, bl > 0. This is a bounded, non-decreasing function of time with lim $\tau \to \infty\ \mu(\tau) = b0$, which is less than infinity, that is, it is a finite failure model.
2. The time between failure of the (i – l)st and the ith failure depends on the time to failure of the (i – l)st.
3. When a failure occurs, the fault, which caused it, is immediately removed and no other faults are introduced.

The data requirements to implement the model is the failure times, τ_i's, of the software system, or the number of faults detected, f, in each period of observation of the software along with the associated lengths l_i of those periods, i = 1, ... n. If data of type 1 are available, the data of the second type can be constructed by first forming a partition of the time period over which the software is observed and then counting up the number of faults that fall in each respective period of the partition. As a consequence, this model can be used for either the time-between-failures data or the number of faults per time period.

EXAMPLE

The Musa and the S-Shaped models were applied to the T1 system. The models were implemented in MathCad software. The details of the implementation are shown below:

The Musa Model

The failure intensity and the mean value functions are:

$$\lambda(\tau) = b0*\ b1*\ e^{-bl*\tau}$$

$$\mu(\tau) = b0*\ (1 - e^{-bl*\tau})$$

Estimation of Parameters

$$b1 = 0.002$$

$$\frac{TNF}{b1} - \frac{TNF*TET}{e^{b1*TET} - 1} - \sum_{j=1}^{TNF} TTF_j = 0 \qquad b1 = 3.841 * 10^{-5}$$

$$b0 = \frac{TNF}{1 - e^{b1*TET}} \qquad b0 = 141.932$$

As shown above, the initial value of b0 needs to be specified. This value is used in the equation for b1 and is iterated till it converges. The failure decay factor is $3.481*10^{-5}$. The value of b1 is then considered for the calculation of b0 which estimates number of faults for the data set. Its value is 141.932.

Calculation of Failure Intensity and Expected Number of Failures

$\lambda(\tau) = b0*\ b1*e^{-b1*\tau}$ $\quad \lambda(0) = 0.005$ $\quad \lambda(TET) = 2.065*10^{-4}$ Failures/CPU Sec

$\mu(\tau) = b0*(1 - e^{-b1*\tau})$ $\quad \mu(0) = 0$ $\quad \mu(TET) = 136$ Failures

The values of the parameters, which are estimated, are then plugged in the equations of failure intensity and mean value function. The failure intensity function λ(TET) gives the failure intensity of the software at the end of the test. Its value for the sample data is $2.065*10^{-4}$. The equation for mean value function gives expected failures at the end of test time, which is 136 failures. Since this is a Poisson type model, the number of failures and number of estimated faults do not need to match at the end of test. This means that there are some undetected faults within the software at the end of test. Fig. 21.2 shows the actual failure intensity and expected number of failures for the given data set.

Graphs for Failure Intensity and Expected Number of Failures

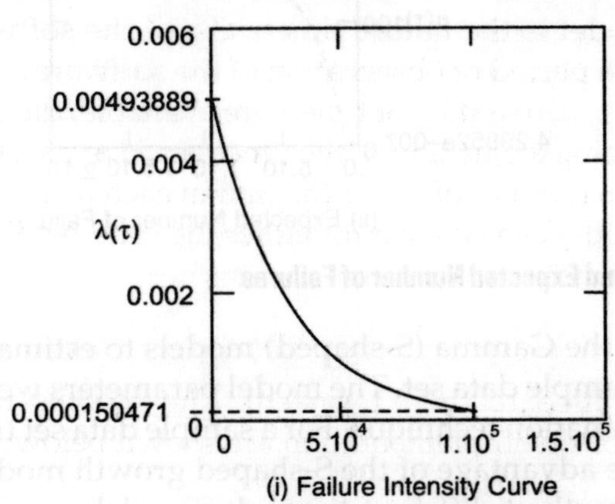

(i) Failure Intensity Curve

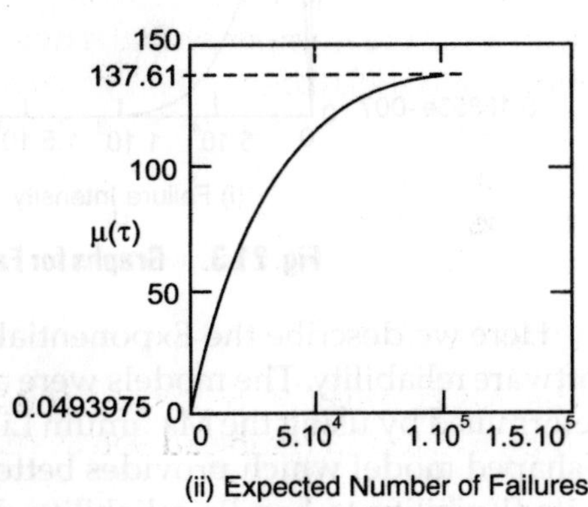

(ii) Expected Number of Failures

Fig. 21.2. Graphs for Failure Intensity and Expected Number of Failures

The S-Shaped Model

The failure intensity and mean value function is given by

$$\lambda(\tau) = b0^* \, b1^{2*} \, \tau^* \, e^{-b1^*\tau}$$

$$\mu(\tau) = b0^* \, [1 - (1 + b1^* \, \tau)^* \, e^{-b1^*\tau}]$$

Estimation of Parameters

$$\frac{2^*TNF}{b1} - \frac{TNF^*TET^*b1^*TET}{e^{b1^*TET} - 1 - b1^*TET} - \sum_{j=1}^{TNF} TTF_j = 0 \qquad b1 = 7.927 * 10^{-5}$$

$$b0 = \frac{TNF}{1 - (1 + b1^*TET) * e^{b1^*TET}} \qquad b0 = 136.82 \text{ Faults}$$

As shown above, the initial value of b1 is specified. This value is used in the equation for b1 and is iterated till it converges. The failure decay factor is $7.927*10^{-5}$. The value of b1 is then considered for the calculation of b0 which estimates number of faults for the data set. Its value is 136.82.

Calculation of Failure Intensity and Expected Number of Failures

$\lambda(\tau) = b0^* \, b1^{2*} \tau^* e^{-b1^*\tau}$ $\lambda(\tau) = 0$ $\lambda(TET) = 5.68 * 10^{-5}$ Failures/CPU Sec

$\mu(\tau) = b0^* \, [(1-(1 + b1^* \, \tau)^* \, e^{-b1^*\tau}]$ $\mu(0) = 0$ $\mu(TET) = 136$

Graphs for Failure Intensity and Expected Number of Failures

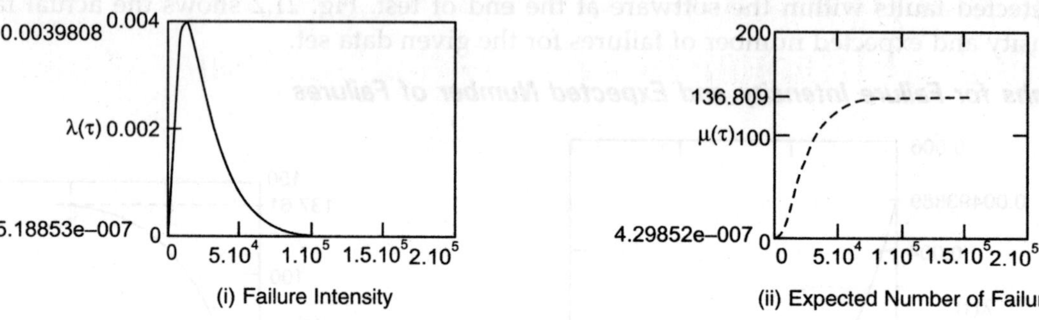

Fig. 21.3. Graphs for Failure Intensity and Expected Number of Failures

Here we describe the Exponential (Musa) and the Gamma (S-shaped) models to estimate software reliability. The models were applied to a sample data set. The model parameters were determined by using the Maximum Likelihood Estimation technique. For a sample data set the S-shaped model which provides better results. The advantage of the S-shaped growth model is its flexibility to handle reliability decay and growth at different times during debugging, something which is common when adding new features to a program. The gamma class model assumes eventual reliability growth, the case when the third parameter becomes one. Under that scenario the gamma model is like the exponential. One can therefore conclude that if there is lack of information on failure data, one could use the gamma class of model vs the exponential class.

22

Updating Manufacturing Technology for Critical Parts Made of Titanium Alloy for Enhancing their Fatigue Characteristics

B.A. Gryaznov, G.V. Tsyban'ov*, O.M. Ivasishin*, P.E. Markovskii*, Yu.S. Nalimov* and O.N. Gerasimchuk**

NTRODUCTION

During their operation, many machine parts and structural elements are subject to the action of the loads varying with time. For this reason, the reliability of their operation and service life depend on the fatigue characteristics of the material used. At the same time, these characteristics can change appreciably while the material undergoes technological operations during the manufacture of a product. This leads to a considerable difference in the calculated and real strength characteristics of a product and to its premature failure.

The present work discusses experimental verification of the influence of technological operations involved in the manufacture of the blades of aircraft Turbo Fan Engine (TFE) fans on the fatigue characteristics of Titanium alloy VT3–1, of which the blades are manufactured, and gives recommendations as to the optimization of the manufacturing methods.

MATERIAL, SPECIMENS, AND EXPERIMENTAL PROCEDURE

Titanium alloy VT3–1 has the following chemical composition (wt. %); Ti – base material; Al – 6.6; Mo – 0.4; Fe – 1.5; Cr – 0.35.

The influence of the production and proposed technologies on the fatigue strength of alloy VT3–1 was checked experimentally on specimens of a rectangular cross-section. They were tested in transverse vibration according to the first flexural mode. A one per cent drop in the resonant frequency of the specimen vibration whereby a fatigue macrocrack initiated in the specimen was taken as its ultimate state.

The microstructure of the alloy at various stages of treatment was studied by the methods of light and transmission electron microscopy. Crystallographic texture of the material was studied and computational programmes developed for the purpose.

*Institute for Strength of Materials, National Academy of Sciences of Ukraine, Kiev, Ukraine.

The specimens and foils for microstructural and texture investigations were prepared using the conventional metallographical methods. The experimental results were processed using programmes written for computers of IBM type.

The variation of the parameters of thermomechanical treatment of the billet for a specimen involved the use of a dual rolling mill and heating in laboratory furnaces resistance type under conditions close to isothermal ones. The final recrystallization annealing was performed in a vacuum furnace at a residual pressure within 10^{-5} Pa.

THE INFLUENCE OF THE MAIN TECHNOLOGICAL OPERATIONS USED IN COMMERCIAL PRODUCTION OF TFE (TURBO FAN ENGINE) FAN BLADES ON THE FATIGUE STRENGTH OF ALLOY VT3–1

The main technological operations involved in the manufacture of blades from billets are:

- Coarse milling of the billet by reducing the machining allowance from 8 mm to 2 – 2.5 mm;
- Fine milling leaving an allowance of 0.5 – 0.8 mm;
- Grinding using abrasive grains (40, 25, 16, 6 μm diameter) and polishing to R_z = 0.63 μm.

The same operations were used to fabricate specimens for fatigue tests from the billets from real blades. Three batches of specimens were prepared whose working sections were treated respectively:

- according to the first of the above operations (batch I);
- according to the first and second operations (batch II);
- according to the first, second, and third operations (batch III).

Figure 22.1 shows the results of fatigue testing of these specimens. From the data presented, in spite of the fact that fatigue of a Titanium alloy is affected appreciably by the degree of surface roughness and the level and distribution of residual stresses in the surface layer, the fatigue limit of the alloy tested during 2.10^7 cycles for all three batches of specimens remains at the level of 550 MPa. As shown by the analysis, some difference in the fatigue endurance curve slopes in the region of limited lives is statistically insignificant.

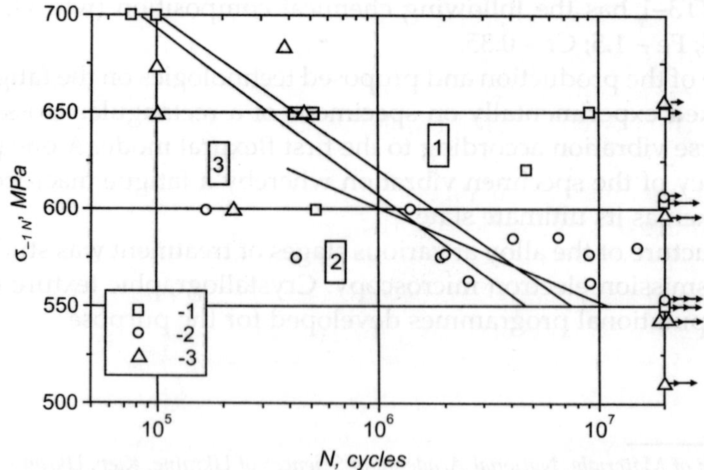

Fig. 22.1. Results of fatigue testing of alloy VT3–1 after various technological operations

Thus, the technological operations and regimes of billets machining used in the manufacture of blades do not reduce their fatigue limit. However, the presence of considerable cyclic life scattering at the stresses above the fatigue limit (e.g., at the stress amplitudes 600 and 650 MPa there are specimens with low lives and specimens which do not fail after 2.10^7 cycles) suggests the availability of unutilized strength margin of the alloy. To verify this assumption, the investigations of alloy VT3–1 were performed at the structural level.

MICROSTRUCTURAL AND TEXTURE ANALYSES OF SPECIMENS OF ALLOY VT3–1 WITH DIFFERENT CYCLIC LIVES

Microstructural and texture analysis were performed on specimens which did not fail at stress amplitudes of 600 and 650 MPa after 2.10^7 load cycles (denoted by 1) and on specimens which had low lives at the same stress amplitudes (denoted by 2).

Typical microstructure of the specimens of type 1 is a homogeneous bimodal structure with a globular 2-phase and the average globule diameter 6–8 μm. The volume fraction of the initial globular α-phase is 60–70%. The rest of the metal is plate-like (α + β)—matrix with the average thickness of α-plates 0.2–0.4 μm.

The specimens of type 2 had some deviations in the structure compared to the specimens of type 1.

These are as follows:

- α-phase had no equiaxial shape with the orientation of the largest dimension of the crystal in one direction;
- An increased size of α-globules;
- Insufficient working through of the alloy for the formation of a bimodal structure.

Such difference in the structure can be explained by the difference in the temperature of thermomechanical treatment and the degree of uniformity of its volume deformation.

The above differences in the structure and the assumed difference in the temperature and deformation conditions of the treatment should also be accompanied by the difference in the crystallographic texture of the phases. Analysis of the specimen texture shows that type 1 specimens exhibit textured β-phase and mixed basic prismatic texture of the β-phase of various intensity, 2-type specimens are characterized by a nontextured β-phase in combination with a texture-free α-phase or with that having basic-type texture of weak intensity.

From the analysis performed it follows that specimens of types 1 and 2 differ appreciably in structure and texture which are not related to specific technological operations involved in mechanical treatment of the specimens but are defined solely by the parameters of thermomechanical treatment of the billets for the TFE fan blades. In addition, the analysis of the fine microstructure of specimens of both types revealed that in specimens of type 1 under cyclic loading the deformations and subsequent initiation of a fatigue crack are localized in the plate-like matrix, whereas α-globules are free from deformation-caused defects. In the specimens of type 2 in the course of fatigue testing the α-phase is subjected to deformation and the initiation of a crack occurs in it. For this reason, obtaining alloy VT3–1 with enhanced fatigue characteristics is associated with the formation of the alloy structure and texture in the blade billet which are optimal from the standpoint of fatigue strength.

DEVELOPMENT OF THE PROCEDURE OF THERMOMECHANICAL TREATMENT OF BILLETS FOR OBTAINING HIGH FATIGUE CHARACTERISTICS OF THE ALLOY

On the basis of the investigations performed the following regimes of thermomechanical treatment of billets have been proposed. To form the most dispersed structure of ($\alpha + \beta$) Titanium alloy, a preliminary quenching is used at a temperature of the single-phase β-region. Water quenching was made at a temperature of 1423 K following isothermal holding for 0.5 hour. In this case, a dispersed martensitic structure was formed in the alloy which, during subsequent deformation on a rolling mill with only 50 per cent reduction (the reduction during one run was 10–12 per cent at the temperatures 1073–1223 K) and final recrystallization annealing (1073–1123 K during 5 hours), provided for obtaining a highly dispersed globular structure. Preliminary thermal treatment of the alloy allows one to eliminate completely the influence of the texture of the initial billet on the processes of texture formation during subsequent rolling.

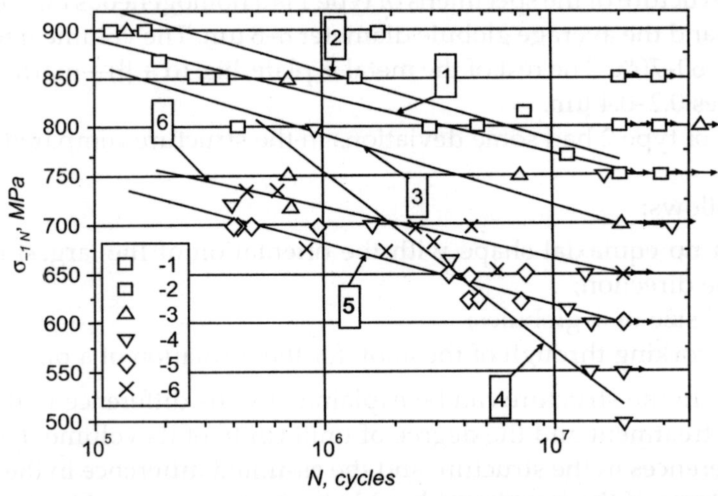

Fig. 22.2. Results of fatigue testing of alloy VT3–1 after various regimes of thermomechanical treatment of a billet for the manufacture of TFE fan blades: according to the developed regime at temperatures of 1073 K (1), 1123 K (2), 1173 K (3), 1223 K (4); "high-speed" treatment (5); and initial state (6)

Figure 22.2 presents the results of testing five batches of specimens of alloy VT3–1, which differ in heating temperature during rolling and were also subjected to unique "high-speed" thermal treatment. As follows from the data presented, the highest fatigue characteristics of the alloy correspond to its rolling temperature 1173 K and "high-speed" thermal treatment. In the former case, this is associated with an increase of the prismatic component in the texture of the basic - prismatic type and in the latter case with the formation of a specific structural state of the alloy with isotropic properties. The fatigue limits of the alloy subjected to the developed methods of thermomechanical treatment exceed those obtained without optimizing its structure and texture (Figs. 22.l and 22.2) by 200 MPa.

- Fatigue characteristics of alloy VT3–1 are little affected by the main operations of the mechanical treatment technology used in the process of full-scale manufacture of TFE fan blades. Of considerable importance here are the structure and texture of the alloy formed at the stage of fabrication of billets.

- Analysis of the microstructure and texture of alloy VT3–1 after fatigue testing revealed that its maximum mechanical properties with a bimodal structure correspond to the fine-dispersed globular α-phase with a mixed basic-prismatic texture of different intensity and thin-plate $(\alpha + \beta)$ matrix with a textured β-phase. The same result with respect to the mechanical properties of the alloy can be obtained with a specific β-structure with isotropic properties.
- Particular parameters of thermomechanical treatment for billets of Titanium alloy VT3–1 for GTE (Gas Turbine Engine), blades which allow one to increase essentially the fatigue characteristics of the alloy and the blades, are identified.

23

A Case for Manufacturing Engineering Education

Robert H. Todd, W. Edward Red*, and Spencer P. Magleby*, Steven Coe***

INTRODUCTION

American industry has awakened to the importance of the manufacturing enterprise and the need for manufacturing education. However, one must ask whether the same urgency has propagated to our educational systems that supply industry with engineers.

In a recent article from *Industry Week*, the following statistics on the influence of manufacturing were noted:

- Manufacturing conducts 90 per cent of all private research and development
- Manufacturing accounts for 74 per cent of all money spent on information technology
- Manufacturing salaries are 89 per cent higher than retailing salaries
- Manufacturing generates about 5 times more jobs than retailing
- Manufacturing represents 80 per cent of all world trade

But how will the professionals so vital to the success of manufacturing be educated and trained? Can traditional educational systems absorb the new integrating business and manufacturing technologies? Is academia sufficiently visionary, or are they encumbered with the same inertia that bowed the knees of U.S. industry for so long? Will new educational structures replace the highly institutionalized systems now in place?

A BRIEF HISTORY OF U.S. ENGINEERING EDUCATION

After World War II engineering education narrowed to include more engineering science activities.

When combined, the 1955 Grinter report on engineering education, the space race, funding for government sponsored research, and the way faculty were evaluated, all catalyzed engineering science activities as the mainstream effort in engineering education for the next 50 years in the United States.

Grinter suggested two thrusts for engineering education. One thrust emphasized engineering science for students who wished to pursue careers in R&D. The other thrust was oriented for students who wished to pursue engineering design activities for practice in industry.

Grinter never intended engineering science to dominate practice-based engineering curricula, but increased government R&D funding in more analytical areas served to focus

*Department of Mechanical Engineering, Brigham Young University, Provo, Utah, U.S.A.

**The Boeing Company, U.S.A.

most engineering programmes towards engineering science. Other countries not caught up in the space race continued to focus on the design and manufacturing side of engineering.

Academia literally plowed under acres of laboratories and shops designed to help students learn how to *apply* engineering fundamentals to the practical problems faced by industry. Evaluation of faculty became based on numbers of research papers published, rather than on evidence of success in preparing graduating students for industry.

The Awakening of American Industry

In the 1970s, American industry began to lose its competitive edge in consumer-product markets as Japan and other foreign countries started to harvest their efforts as a result of their emphasis on design and manufacturing activities.

The U.S. clearly had the upper edge in the invention of new technologies, particularly in aerospace, computers, and defence. However, in consumer-product development and manufacturing, Japan and other foreign countries learned how to make products of higher quality and lower cost. The U.S. was busy with the space race and the Cold War, and America's trade deficit soared.

Frederick W. Taylor, considered the father of American manufacturing, created the mass production model, based on the premise that complex jobs should be broken down into repetitive, routine tasks. Workers could be trained to perform these tasks reliably: "Design and planning were accomplished by a group of educated managers with whom total decision-making power rested."

Created at the turn of the 20th century, the Taylor model is now in direct conflict with the changing technologies and empowerment cultures needed for modern business and manufacturing enterprises. Modern enterprises confront daily a dynamic product market, where product changes are measured in weeks and months, rather than years, and where information is available to many.

Titone used a *New York Times* article to estimate that only 5 per cent of America's corporates have made the transition to a more open process of enterprise integration and shared decision-making, processes that are fundamental to Japanees industry. It seems likely that the lack of young engineers trained in the new enterprise technologies may contribute to the entrenchment of Taylor's model in U.S. industry.

Table 23.1 is a partial listing of enterprise technologies that industry is struggling to adopt. In general, conventional engineering curricula have not adapted to these new needs.

Review of the ABET (Accreditation Board for Engineering and Technology) Criteria as Another Statement of Industrial Needs

Since manufacturing can occur in a number of traditional engineering disciplines, such as Mechanical or Electrical Engineering, some suggest that manufacturing cannot be viable as a program of study in higher education.

The reality is that engineering disciplines (or fields) distinguish themselves by application areas, by the language and terminology unique in these areas, by specific and collected application data, and marginally by principles that may be unique to an area. For example, controls are taught in EE (Electrical Engineering), ME (Mechanical Engineering) and ChemE (Chemical Engineering). Many of the control principles are found in fundamental physics and chemistry. The state vector and differential control mathematics can be found in fundamental and advanced math courses. Yet, the controls domain (or application areas) will vary by engineering program, e.g. process control in ChemE, electrical/electronics control in EE, and mechanism control in ME.

Reviewing the 1998–1999 ABET curricula criteria for four engineering programmes, ME, CE (Civil Engineering), EE and MFE (Manufacturing Engineering) reveals that, except for manufacturing engineering, the programmes criteria provide little distinction from one another. In contrast, the published ABET criteria for Manufacturing Engineering clearly delineates four distinct areas:

- Materials and manufacturing processes,
- Process, assembly, and product engineering,
- Manufacturing competitiveness, and
- Manufacturing integration, and systems design.

These areas, seemingly unique, appear to fall outside the domains of traditional CE, EE, and ME.

Traditional Mechanical Engineering Programmes

In the 1980s, American industry clearly voiced concerns about the preparation of engineering students, but these concerns often fell on deaf ears. The emergence of engineering technology programmes, started in the 1960s in the U.S. helped fill the educational void left when engineering science curricula replaced practice-oriented curricula. However, these programmes have not received acceptance by some as 'engineering programmes'. Jerry Junkins, past Chairman and CEO of Texas Instruments, put it this way: "Most engineering jobs involve design and practice, not theory and research."

Conventional engineering programmes and departments have been impacted by increased manufacturing-related research, improved placement opportunities for manufacturing engineers, and students more interested in practice. Some common reactions have brought about the introduction of new manufacturing engineering departments. Changes in the names of departments include manufacturing in the title, a few new manufacturing elective courses, and manufacturing programmes offered as an interdisciplinary effort of several departments, or as a specialized graduate programme.

Of the traditional engineering curricula mechanical engineering is the most closely associated with manufacturing. Some ME programmes have worked to include aspects of manufacturing in their curriculum. This has been especially true in graduate studies and research related to manufacturing processes. The reality is, however, that it is difficult for an undergraduate ME curriculum to stretch its course capacity to include a comprehensive range of manufacturing principles and applications.

The typical mechanical engineering curriculum is shown in Table 23.2. Its discussion about which courses could be modified or replaced to emphasize manufacturing would bring heated discussions in any faculty meeting.

If we consider that we are trying to address the needs of manufacturing enterprises, which we are trying to serve, should we not ask their opinion on this issue? In the following section, we present the results of a survey on manufacturing education, administered to industry.

Table 23.1 New Business and Manufacturing Enterprise Technologies

Technology	*Description*	*U.S. Industry Status*	*Educational Preparation*
Product teams	Concurrent engineering, colocated, multi-disciplinary, geographically separated, culturally diverse, multi-language, international.	Strong progress in the last 10 years driven by competition for shorter product cycles.	Engineering programmes traditionally separated by discipline. Undergraduate capstone programmes may be multidisciplinary.
Lean manufacturing (optimal control of known manufacturing environments)	Cellular manufacturing, JIT, Kaizan, Kanban, manufacturing process and material awareness	Enterprises that want to stay competitive have made changes and implemented a number of these concepts.	Few of these topics have been introduced in graduate courses, almost none in undergraduate courses.
Agile manufacturing (responsive, flexible manufacturing environments to provide a multiplicity of finished products)	Customer directed, dynamic and flexible production systems, simulation and virtual environments.	Enterprises with short new product introduction cycles have responded to these new methods best.	Few schools and departments have educational activities and facilities designed to prepare students in these concepts.
Business enterprise	Project management (scheduling, cost control, resource planning, estimating performance visibility); inventory management, shop floor control, value chain awareness; make or buy decisions, supply chain management.	Principles successfully introduced to large and small companies, competition being the driver. Some larger companies have had more difficulty in getting "the water to the end of the row" than small to medium-sized progressive companies.	Concepts often taught in business school courses but rarely in engineering programmes.
Automation	Robotics, CAM, CAPP, vision and sensor systems, networking open-systems.	Implemented more intelligently with improved ROI (Return on Investment).	Few engineering schools introduced these topics, typically at graduate level.
Information Technology	Integrated business and data systems, product definition, data management, internet, company/supplier networks.	Best implementation of new technologies has occurred here; companies want more.	Business schools have been the most responsive in introducing these concepts.

Table 23.2 Typical Courses in Mechanical Engineering Curriculum (Not including math and science pre-requisite core)

Subject Areas	*Number or Courses*
Economics	1
Statistics	1
Measurement, Experimentation	2
Systems, Control	1–2
Mechanics	2
Materials	2
CAE, CAD, Product Design	1–2
Computing Techniques	1
Thermodynamics, Fluids, Energy	2–4
Design (Capstone, Machine Design)	2–3
Manufacturing Processes	1
Technical Electives	2–3

What does the Industry Need From Academia?

We completed a survey of 120 engineers, managers and upper level executives from 44 different companies. Four questions were posed, three of which are presented here.

1. *Please indicate the most important technical or other challenges that you expect new engineering graduates to face in your company in the next five years.*

Response
Working collaboratively; systems engineering, design for manufacturing, communication, dealing with change, lean manufacturing. Understanding manufacturing processes. (areas most marked with a 5, i.e. highest importance.)

Observation
The word "manufacturing" dominates the responses.

2. *In your opinion, how strong is the need for undergraduate manufacturing engineering programmes in the United States?*

Response

No need 1	2	3	4	*Strong need* 5
0.0%	1.7%	4.2%	31.9%	62.2%
			94.1%	

Observation
Overwhelming support for undergraduate manufacturing engineering programmes.

3. Universities have tried a number of different approaches to get more manufacturing to their students. These include: (a) Adding one or two elective manufacturing classes to traditional engineering majors, (b) Creating an option or emphasis in manufacturing within traditional engineering majors, (c) Creating a manufacturing engineering degree, (d) Treating manufacturing primarily as a graduate area with no special undergraduate preparation. Please prioritize each alternative from 1 (highest) to 4 (lowest). Use each number only once.

Response

Approach	*Highest* 1	2	3	*Lowest* 4
(a) Add classes	17.7%	18.6%	44.2%	19.5%
(b) Option	29.9%	46.2%	13.7%	10.2%
(c) Degree	51.3%	19.3%	20.2%	9.2%
(d) Grad. area	11.2%	12.9%	17.2%	58.7%

Observations

Strongest support for option or full manufacturing programmes. Less support for adding manufacturing classes to existing curricula or adding graduate manufacturing programmes.

Note: The 120 respondents to the survey from the 44 different companies represented the following functional job titles:

(a) CEO/Presidents, Vice Presidents or General Managers—12
(b) Managers—39
(c) Manufacturing engineers—14
(d) Other engineers—17
(e) Others—38

57.4 per cent responded "yes" to the question whether or not they, as individuals, were in a position to recommend new engineering graduates for hire.

Traditional ME courses are oriented towards performance-based analysis sprinkled with some design. If manufacturing is to be included as part of a an ME programme, there are a number of possible approaches. Noted below are four possibilities.

1. Increase the awareness of manufacturing through modifications of existing ME courses. This could be done without a complete overhaul of the curriculum. While this approach does bring some additional emphasis to manufacturing most of the faculties that teach traditional ME subjects do not have experience in manufacturing applications. This approach would likely leave many important manufacturing topics, needed by industry, unavailable to students.

2. Provide courses in manufacturing that can be taken as electives to fulfill ME requirements. In most curricula, these courses would be considered electives with some restrictions placed on them. Examples of such courses might include manufacturing processes, lean manufacturing, quality assurance, manufacturing competitiveness or other subjects noted in Table 23.1.

3. Reduce traditional ME content so that there is room for a cohesive group of manufacturing courses as a manufacturing emphasis area or option. Sorting out which courses would be eliminated would be an arduous task for any faculty.

4. Develop separate manufacturing engineering programmes whose curricula follows the principles, fundamentals and applications needed by industry and outlined by ABET. The reality is that it is extremely difficult for existing engineering programmes to change their curricula to incorporate manufacturing practice as suggested in options 1–3. Recently, a curriculum study of undergraduate mechanical engineering programmes in the Big Ten (Michigan, Illinois, etc.) was assembled. The results of the study showed that of the required 130 hours, an average of less than three credits were specifically labelled as "manufacturing".

Future Directions in Engineering Education

ABET, with input from 22 engineering societies, recently defined an outcome-based approach to accreditation in an effort to improve the accountability and relevancy of engineering education. "The ABET 2000 shift towards outcome-based education is analogous to the total quality movement in business and manufacturing."

Faculty may not yet understand the broad competitive implications of this educational transformation. The independence to define and protect curricula in the past is being replaced by a methodology for comparative programme evaluation. When we factor the information technology revolution into the outcome-assessment paradigm, academic competition will surely awaken prospective students to the best provider and medium of choice.

As stated by Edward Goldberg, "A significant proportion of "physical face to face" education will be replaced by Web delivered, electronic learning opportunities. Educational institutions which do not transform themselves into "electronic learning providers" will find themselves losing their students not only to current competitors, but to new forms of institutions as well. Those which realign their practices with the new technologies will reap substantial benefits. As transformations take place, the major beneficiaries will appropriately be students."

The Internet, and its easy to use multimedia part, will profoundly impact the way we live, work and learn. Bill Gates is correct when in his book, *The Road Ahead* (Viking, 1995), he draws an analogy to the impact of the Gutenberg press on the Middle Ages. Published outcome-based curricula will be much easier to adapt or duplicate than existing curricula, now confined to an instructor's notebook. Concepts, such as **net-design** and **net-control,** introduce the possibility of not only distance learning, but also *distance practice*.

Net-design is the engagement of experts and design tools over the Internet. In this mode, students interact across the Internet with team members and design moderators (instructors, assistants, design experts), using distributed design tools to view and modify design instances.

Net-control is the similar engagement of modern, but distributed, equipment such as robots, CNC machines, vision systems, etc.

It is the *practice* component that has protected the curricula of professional engineering programmes from outside competition. The possibility of distance practice over the Internet may provide new scenarios for practice education, permitting a learner, more economically efficient class of educational providers.

Manufacturing Engineering Education in Other Countries

About a year ago, a special workshop, "Excellence in Global Manufacturing Education", sponsored by the National Science Foundation, was held at Arizona State University, U.S.A. There were about 30 invitees at the meeting representing industry, academia and government from 10 different nations representing North America, Latin America, the Far East, India and Europe.

At one of the meetings a faculty member, Poul H.K. Hansen from Aalborg University, Denmark, was asked what percentage of students in his university were majoring in traditional engineering disciplines such as mechanical or electrical engineering versus manufacturing engineering.

He explained that there were 3-4 manufacturing students for every mechanical or electrical engineering student. This was the case because of a decision made 15 years ago in Denmark. He further explained that his country decided that if they wanted to compete in the global economies of manufactured goods, they must put more emphasis on manufacturing engineering.

Students, he said, simply know where the jobs are today and where their future is. He also explained that most practicing engineers learn fairly quickly that their company must have a set of products that will compete in the market place. To do this, it simply takes, proportionally, far more engineers to get new and better manufacturing processes and process machines developed, designed and in control throughout the life of a product than it does to design the product itself. For an enterprise to compete in a global economy there are simply more engineering jobs related to manufacturing than product analysis or design.

An employee from NIST (National Institute for Standards and Technology), U.S.A., recently told one of the authors that, in the U.S. approximately 50 per cent of all engineers end up working in manufacturing, yet only 5 per cent of all engineers have had any meaningful educational training directly related to the challenges of manufacturing. This statistic, true or even near true, should stimulate changes in all engineering curricula.

Japan's undergraduate educational approach for engineering is different from both the U.S. and Europe. With the practice of lifetime employment in Japan, industry assumes the responsibility of educating their engineers for engineering science activities *and* industry practice activities. Japan's manufacturing capability results from a culture that influences cooperation and dedication, and offers lifetime training of employees. As a result, academia seems to be robbed of its accountability for preparing practicing engineers or engineering scientists in Japan.

In Europe, universities never separated engineering science activities from engineering practice activities, unlike the U.S. There are no engineering technology programmes since engineering science and engineering practice activities are well integrated in their engineering programmes.

Whereas Japanese industry has assumed the accountability for practical and lifetime training of its engineers, Europe has used a strong partnership between academia and industry to balance practice with theory. In the U.S., the educational accountability between industry and academia is still not well defined.

The growth in information technologies may cloud the accountability relationship even further.

Growth in U.S. Manufacturing Engineering Education Programmes

The number of manufacturing education programmes in the United States is growing substantially. In 1990, there were only three accredited BS manufacturing engineering programmes in the United States and 19 four-year accredited manufacturing engineering technology programmes. In December 1997, there were 16 accredited undergraduate manufacturing engineering programmes and 37 four-year accredited manufacturing engineering technology programmes. Four new MFE (Manufacturing Engineering) programmes applied for accreditation in the fall of 1998.

One of the authors conducted an internet survey to find more than 60 *universities* advertising some form of U.S. manufacturing engineering programme (not counting technology programmes or the rapidly proliferating international manufacturing programmes), either undergraduate, graduate, accredited or unaccredited.

These new arrivals include Stanford, Berkeley, Michigan, and other well-known schools. This growth rate shows a new manufacturing sensitivity in educational institutions.

Forty of the fifty states now offer professional registration in Manufacturing Engineering. This compares with only 32 for aeronautical/aerospace engineering and 36 for structural engineering. Forty-nine states offer professional registration for chemical, electrical, mechanical and civil engineering.

We have reviewed the history of engineering education, examined the content and directions of current mechanical engineering programmes, and have presented data from engineering professionals that describe their desire for more manufacturing emphasis in engineering education. We have also examined tools that may provide improved education in manufacturing.

We have presented this information in an effort to foster review of educational content in traditional engineering disciplines, especially the related discipline of mechanical engineering. Our analysis of the information and the current industrial environment indicate that manufacturing represents a strategic direction and opportunity for engineering education to pursue.

24

Tribology in Product Design and Manufacture

*R. Williams**

INTRODUCTION

In any manufacturing situation, two important elements interact if the product being marketed is to have market appeal and be profit making. They are: the design concept of the product and the capability of the machine(s) employed to make the product. Both the elements have common fundamental features—they must perform their allotted task and do it efficiently, shortcomings in either area will result in an unsaleable product. To achieve this objective, due attention must be paid to the tribological aspects involved.

Superficially, tribology appears to some as a purely academic exercise while to others, it is a simple and mundane topic. In practice, when properly considered and applied, it has many diverse facets requiring expert and experienced engineers to achieve full benefits. The objective of this paper is to provoke a greater awareness of the science of tribology, to provide reference to "tools" for its application and to highlight benefits that can be achieved with "Case Histories from a Tribologists Case Book". The range of topics embraced by the science of tribology is extremely wide. Therefore, the scope of this paper has been limited to a discussion on the various types of bearing and their application.

Tribology is the science and technology of interacting surfaces in relative motion. It involves friction, wear and lubrication. Friction is in the domain of the engineer; wear, the metallurgist and lubrication, the chemist. It, therefore, follows that tribology is a multi-disciplinary science and those that practice it should have some background in all these areas if he, or she, is to make a serious impact on tribological situations. Generalising, the largest proportion of tribological attention is devoted to machines, but it is by no means limited to machines.

A machine may be defined as an apparatus having several moving elements, each with a definite function in an organised system. It follows that all machines will have a number of interacting surfaces in relative motion, i.e. bearings, each requiring a long efficient life at minimum installation and operating cost. From this it may be concluded that the greater number of triological problems occur with machines.

The product can be anything: another machine; a simple cooking utensil; a garment or purely an ornament. Nevertheless, in all cases, some aspect will require tribological attention. Typically, a cooking utensil must resist corrosive, erosive and abrasive wear; a garment needs

*Consultant Engineer, R&P Design, UK.

acceptable friction (feel) and good wear resistance; an ornament, good finish and so on. Neglect of such features can lead to tribological problems and/or an unacceptable product.

In the author's view, most tribological problems that do arise originate in the design office and it is in this domain, in particular, that a more determined approach to the subject of tribology is required if many of the troublesome symptoms found in service are to be avoided.

PART 1: BEARING SELECTION

The Design Approach to Bearings

The basic requirements of a bearing are: to support the load, permit motion, and to have an adequate life. By including some additional features and developing them further, a typical bearing design specification can be derived from the following features:

Load - Motion - Speed - Location - Accuracy - Environment - Life - Cost (initial and operating). Additional features may become necessary, dependant on special applications, i.e. aerospace, nuclear, marine etc., but the prime objective is always to define the functional requirements before looking for a bearing and, indeed, to determine whether a bearing is really necessary.

Tribological Solutions

As already stated, tribology deals with the friction and wear that occurs between moving surfaces when a load is applied across the interface. If the resultant effect is to be acceptable to the design, a first consideration will be the choice of the rubbing materials (Fig. 24.1(a)). In tribology, there must always be at least two materials involved. If the choice of materials does not yield an adequate solution then surface treatment(s) or chemically bonded films, i.e. MoS_2, dry film lubricants, may be considered (Fig. 24.1(b)). To minimise friction, and or wear, the surfaces may be completely separated by means of rolling elements (Fig. 24.1(c)) or, alternatively with a film of fluid. The prime requirement of a fluid film is to have a pressure sufficient to support the applied load. This pressure may be externally applied, the so called hydrostatic bearing (Fig. 24.1(d)) or be self generated from the motion of the bearing, the hydrodynamic bearing (Fig. 24.1(e)). In the latter three options, it is essential that a "lubricant" is present, which may have to be isolated from the environment, introducing a need for seals—another area for tribological attention.

The surfaces may be separated by means of a magnetic field, passive or active (Fig. 24.1(f)). Another alternative is to simply introduce an elastomer, such as a rubber bush, or a system of metallic strips, the flexural pivot (Figs. 24.1(g) and (h)), between the moving elements. In these three systems, the tribological problem is, in effect, "designed out" but the latter two only allow a limited deflection and are therefore confined to oscillating movements.

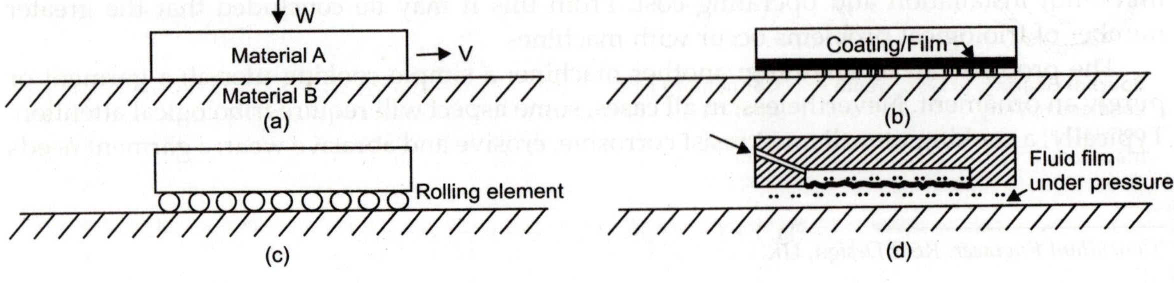

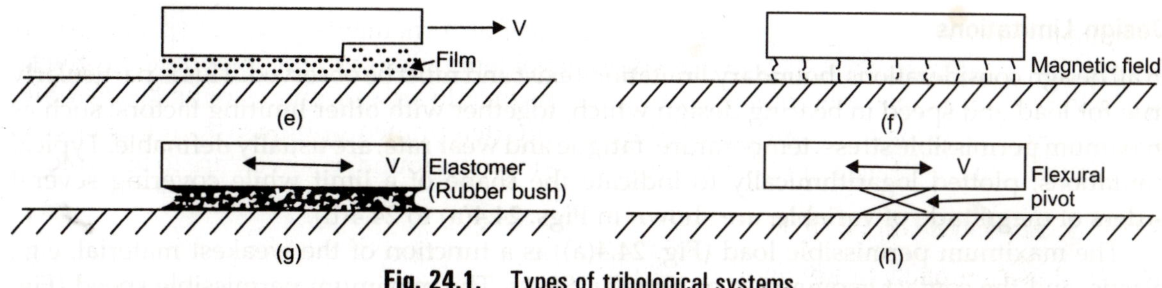

Fig. 24.1. Types of tribological systems

Load and Speed

In considering load and speed it is readily apparent that both have magnitude and direction, each of which may be constant or variable. This results in four classes of loading (Fig. 24.2(a)), whilst speed can have up to sixteen classes (Fig. 24.2(b)) when a combination of rolling with sliding; e.g., as occurs between gear teeth, and impact with sliding; e.g., as occurs in mineral conveying, are considered. Typical examples of the sixty-four possible load/speed combinations are shown in Fig. 24.3. Although no attempt is made to define the range and pattern of load and speed combinations and their variations, it can be seen that these two variables alone can lead to a very wide range of tribological situations. Conveniently, in a limited discussion considering bearings only, roll/slide and impact/slide may, in general, be ignored, reducing the number of classes to sixteen combinations of load and speed. Furthermore, most variable load/speed situation designs can be based on constant load and constant speed and still yield acceptable solutions, again reducing the number of classifications for consideration.

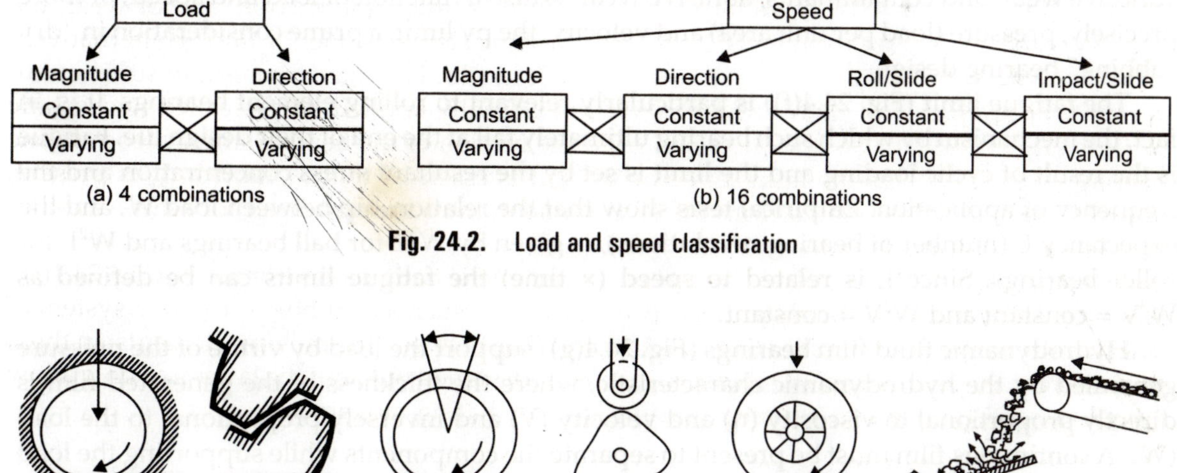

Fig. 24.2. Load and speed classification

Fig. 24.3. Types of load and speed variations

Design Limitations

In all design considerations, boundary limitations exist and must be observed. This is particularly true for load and speed in bearing design which, together with other limiting factors, such as maximum permissible stress, temperature, fatigue and wear rate, are usually definable. Typical limitations, plotted logarithmically to indicate the shape of a limit while covering several orders of magnitude of variable, are shown in Figs. 24.4(a) to 24.4(h).

The maximum permissible load (Fig. 24.4(a)) is a function of the weakest material, e.g., plastic, and the contact geometry, e.g. *Hertzian Stress*. The maximum permissible speed (Fig. 24.4(b)) can be limited by inertia characteristics, since all components in motion are subject to acceleration fields which give rise to enhanced forces due to their inertia. Typically, a rotating shaft may burst due to centrifugal stress or an oscillating linkage may impose bearing loads far in excess of those determined by component weight or the force to be transmitted. Yet again, rolling element bearings set up significant centrifugal and gyroscopic forces as speed increases which can result in failure due to ball skidding and/or burst cages. In addition, all rotating systems can have speed limitations imposed to avoid the onset of resonant vibrations, i.e. shaft or bearing whirl.

The lower end of permissible speed (Fig. 24.4(c)) can also be a limiting factor, where stick/slip in rubbing bearings and inadequate film thickness in hydrodynamic bearings gives rise to instabilities and early failure.

The thermal balance for the removal of friction generated heat from the bearing location defines a further limit on load and speed (Fig. 24.4(d)).

Whilst the acceptable wear limit (Fig. 24.4(e)) is mainly affected by materials combinations, adhesive wear, and contaminants, abrasive wear, is also a function of load and speed, or more precisely, pressure (load per unit area) and velocity, the pv limit, a prime consideration in "dry rubbing" bearing design.

The fatigue limit (Fig. 24.4(f)) is particularly relevant to rolling element bearings. It is, in fact, the mechanism by which such bearing ultimately fail at the end of their design life. Fatigue is the result of cyclic loading and the limit is set by the resultant stress concentration and the frequency of application. Empirical tests show that the relationship between load W, and life expectancy L (number of bearing revolutions), is given by W^3L for ball bearings and W^2L for roller bearings. Since L is related to speed (× time) the fatigue limits can be defined as W^3V = constant and W^2V = constant.

Hydrodynamic fluid film bearings (Fig. 24.4(g)) support the load by virtue of the pressure generated by the hydrodynamic characteristic, where the thickness of the generated film is directly proportional to viscosity (η) and velocity (V) and inversely proportional to the load (W). A continuous film must be present to separate the components while supporting the load at a given speed with the appropriate lubricant. The limit of operation can therefore be defined by the expression, V/W = constant. As speed (velocity) increases, viscous shear will raise the temperature of the fluid and therefore reduce its viscosity causing a departure from linearity as shown.

In the hydrostatic bearing (Fig. 24.4(h)) the fluid film thickness is dictated by an external pumping system which, in turn, defines a minimum load carrying capacity irrespective of speed. Again, increased frictional heat will cause a departure from linearity at the higher

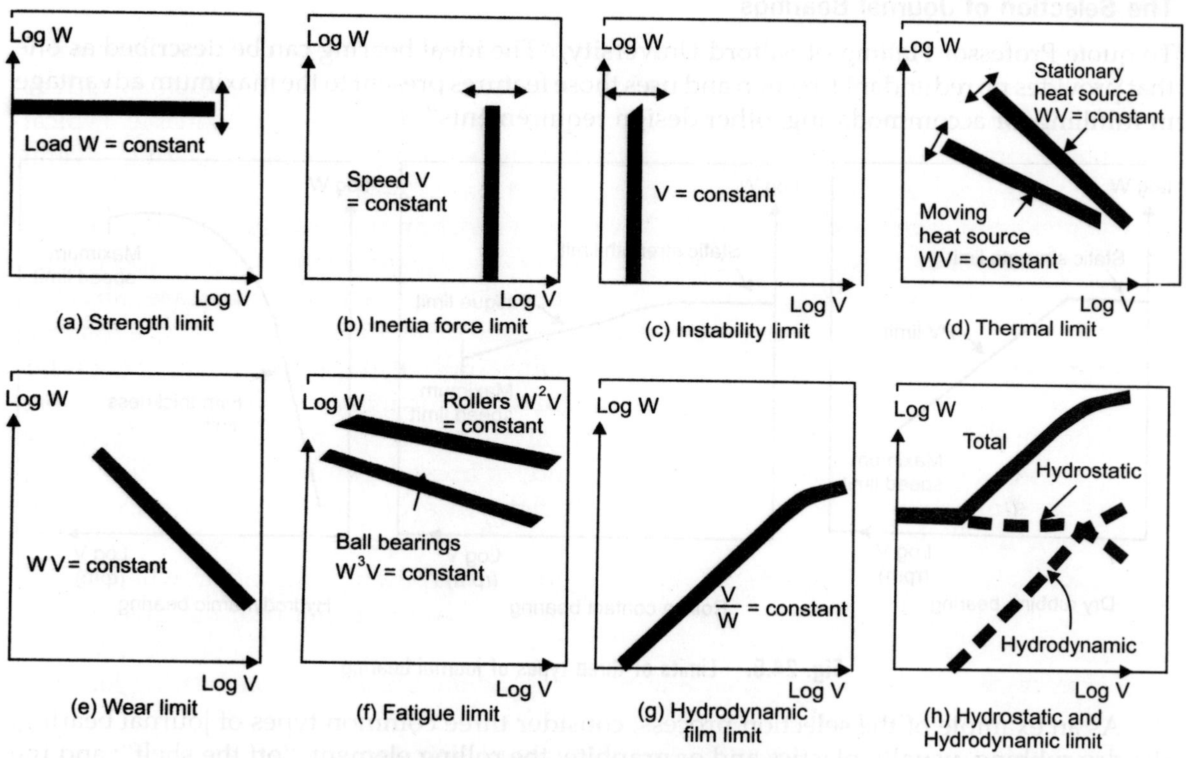

Fig. 24.4. The nature of tribological limits

speeds. However, with increased speed hydrodynamic effects enhance the total load carrying capacity which is then the sum of the two characteristics as shown in Fig. 24.4(h).

The presentations in Fig. 24.4 provide no numerical values for the various limits shown as these will depend on the many factors applied to the design, such as materials, geometry, fluid, etc.

The Environment

The nature of the environment in which a tribological device (a bearing) is to operate has a significant influence on its proper functioning. Situations, such as high vacuum, high or low temperature, radiation fields (nuclear), aggressive process fluids (acids, sodium) etc. present special problems which will greatly affect the choice of a tribological solution. Atmospheric conditions also present a wide range in the operating environment which can affect the performance of most types of bearing, e.g., temperature, humidity and airborne dust. High altitude (aircraft) and outer space (satellites) combine problems of extremes of variable pressure and temperature where lubricants can rapidly evaporate and protective oxide films stripped from component. Process industries invoke the separation of lubricants and process fluids, in some cases absolutely, and recourse is taken to using the process fluid as a lubricant. Typical examples are those of the bearings in sodium pumps used in nuclear fast reactor technology and in fluorine compressors (gas bearings) in the chemical industry. The ability to apply the required expertise to the problems involved in using process fluid as a lubricant is very much in the hands of the experienced engineer and tribologist.

The Selection of Journal Bearings

To quote Professor Halling of Salford University, "The ideal bearing can be described as one that provides no redundant function and uses those features present to the maximum advantage in fulfilling, or accommodating, other design requirements".

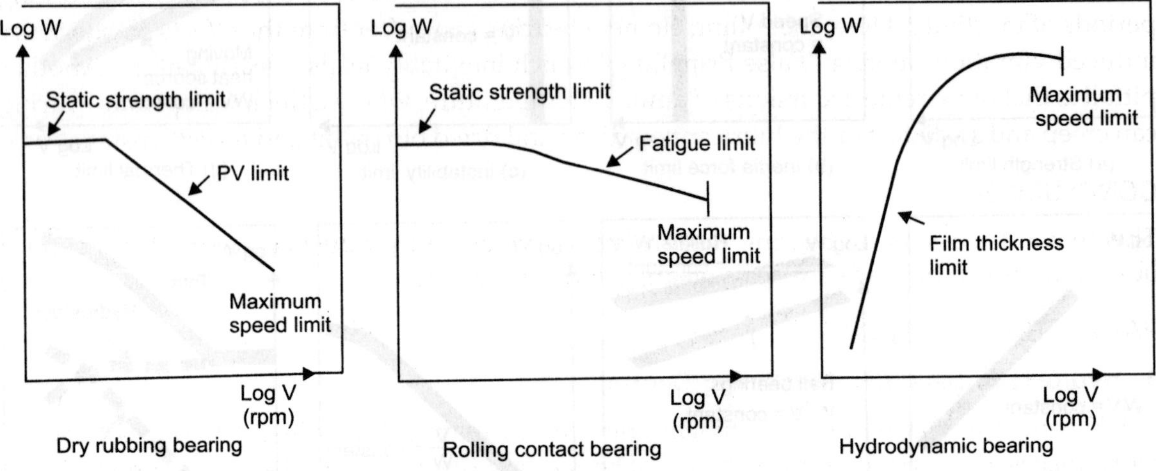

Fig. 24.5. Limits of three types of journal bearing

As an example of the selection process, consider three common types of journal bearing: the dry rubbing, usually plastics and or graphite; the rolling element, "off the shelf"; and the hydrodynamic, generally custom designed and built. Referring to the constraint limits shown in Fig. 24.4 it can be seen that the basic characteristics for each of these bearing types will be of the form shown in Fig. 24.5.

Rolling element bearings are advantageous (in load carrying capacity) for shaft sizes below 25 mm., in the speed range 800 to 3000 rev/min. This, together with the ease of selection and installation, can explain why they are so often used. Hydrodynamic bearings of larger diameter in the same speed range have still higher load carrying capacity. Sintered metal bearings are a very important category of bearings, frequently used in small electric motors and home appliances where its low cost and "fit and forget" characteristics are invaluable.

Bearing design is, of course, an iterative process and design charts, as in Tribology Handbooks, are helpful in the selection of the right bearings. However, more detailed calculations will still be required to ensure that the selection is "fit for purpose".

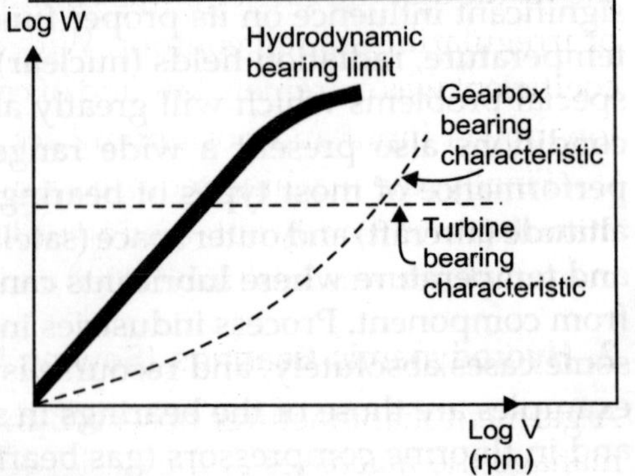

Fig. 24.6. Matching of tribological solutions

Fit for Purpose

Matching the characteristics of a bearing to the requisite duty is illustrated by referring to Fig. 24.6, where the load/speed characteristics of a large gearbox bearing might be of the form shown (load building up with increasing speed), an hydrodynamic bearing is a good match for this duty. Similarly, the heavy rotor of a steam turbine (where load is substantially constant) is better served by an hydrostatic bearing, although, with prudent design, the hydrostatic function could be employed as an auxiliary jacking system, purely for start up and stopping, and be turned off when a certain speed was achieved.

Care must also be taken to cover "off duty" conditions, such as transportation and long periods of inactivity. Here, the rolling element bearing can suffer from the effects of externally induced vibration causing "False Brinelling" which inevitably leads to early failure. Another pitfall is that of extreme excursions of ambient temperature, where a thick wall plastic bearing can creep and so distort or the lubricant oxidised and dried out in "fit and forget" installations.

CONCLUSION

The above discussion on bearing design only serve as an introduction to tribology, similar observations can be made for seals, gears, power trains, etc.

PART II: CASE HISTORIES

1. Hydrostatic Bearings (Machining Centre)

A UK company was to supply a machining centre capable of handling 400 tonnes (weight) work-pieces with a positioning accuracy of better than 10 μm. The work-piece, supported on a two piece carriage weighing 200 tonnes, traversed over a 20 m. stroke under a gantry on which the various machining heads were mounted (Fig. 24.7(a)). Previous production machines were of much lower capacity and used a proprietary lubricated overlay on the work carriage guide rails to minimise friction (Fig. 24.7(b)). This system was totally inadequate for the new machine as the high friction and "stick/slip" between carriage and rails defied any attempt at precise positioning and absorbed large amount of power during positioning.

A tribology consultant was invited to investigate the bearing problem early in the design stage. After a careful study, all the available "off the shelf" anti-friction material combinations were eliminated and the agreed viable solution was to use hydrostatic bearings to support the carriage and work-piece. The design consisted of a multiple of oil fed pockets. (Fig. 24.7(c)), each supplied by separate constant volume output pumps. This ensured that pocket pressure in the low loaded areas of the carriage dropped and that in highly loaded areas increased, with the result that the carriage was continuously "floated" on oil irrespective of load distribution or irregularities in rail contour. With no direct contact between carriage and rails the friction coefficient was extremely low and of constant value, static and dynamic. This system demanded low power input and positioning accuracy was easily achieved.

Finally, the machine required a minimum of development and was so successful that the manufacturer readily obtained an order for another machine. The cost of the consultancy to the manufacturer was less than £3000 and each machine was sold for in excess of £1,000,000 each.

2. Hydrodynamic Bearings (Sewing Machine)

A garment manufacturer observed that his overedge sewing machine operators could easily manage the machines at the designed stitching rate of 4000 spm (stitches per minute), (4000 rpm). To improve productivity he changed the drive pulleys and increased the stitching rate by

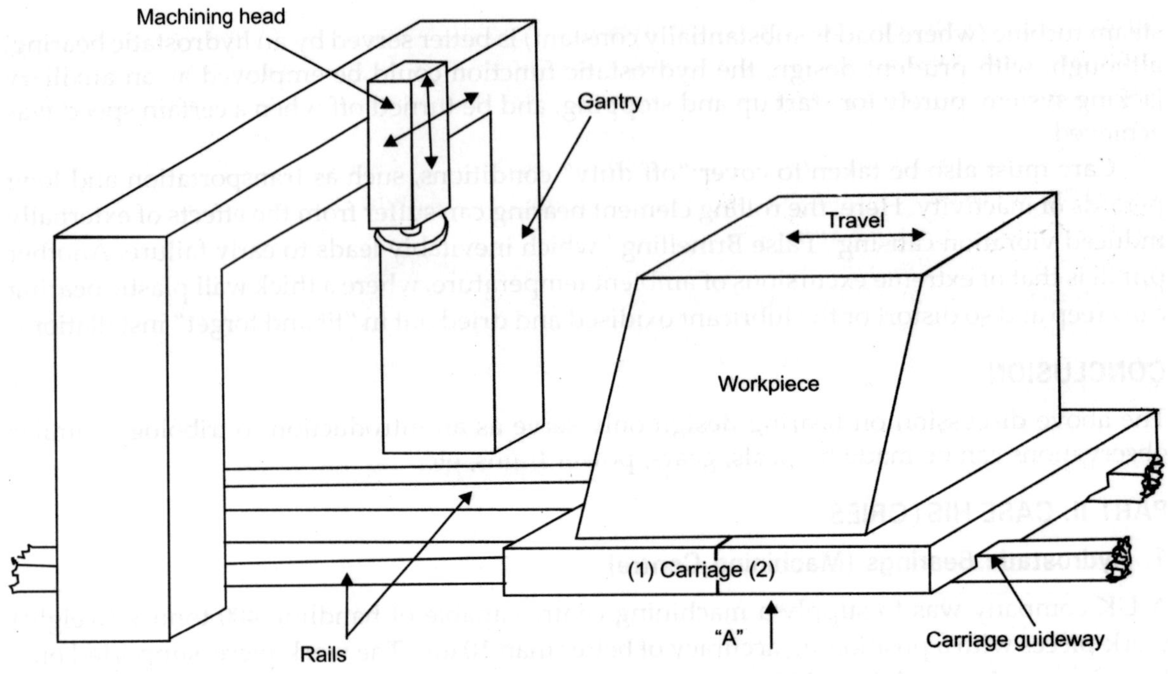

(a) Diagrammatic view of machining centre

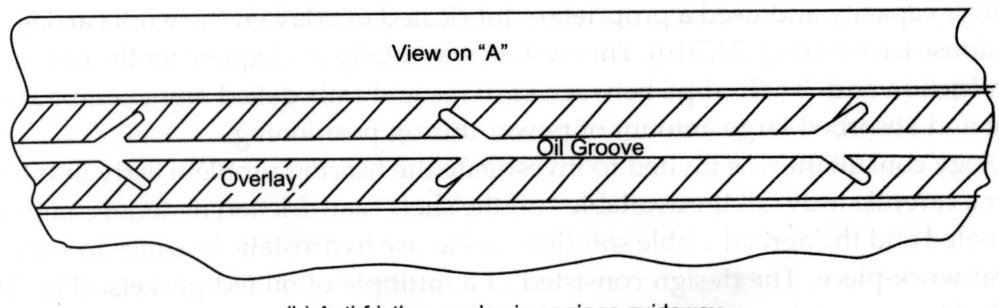

(b) Anti-friction overlay in carriage guideway

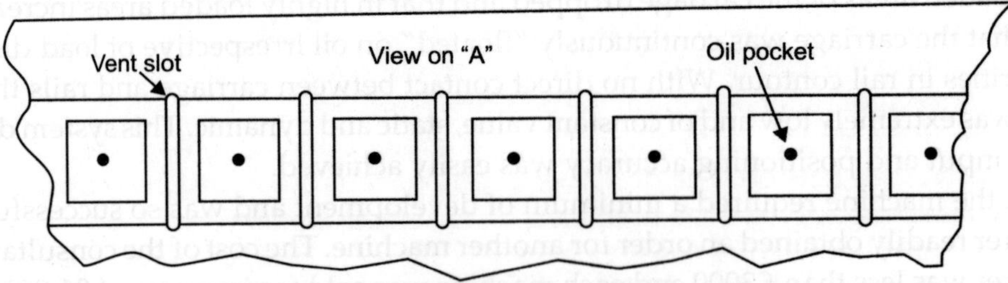

(c) Hydrostatic bearing pockets in carriage guideway

Fig. 24.7

10 per cent. Again, he found that the machine operators had no problem and there is no increase in reject work. Following this, and in conjunction with the machine manufacturer, the machine stitch rate was progressively raised to 9000 spm and still the operators were able to manage the machines efficiently. However, any further speed increase was not possible because the *machine* failure rate had become unacceptable. They diagnosed the problem as being due to a bearing/lubrication failure and invited a tribology consultant to analyse the effect of increased speed and to identify the problem.

Overedge sewing machines eliminate the need to overfold fabric in final sewing by using three needles to weave thread over the raw edges to resist fraying in service. Typical examples are: garment inner seams and woolen blanket edges. The needles are driven via a system of links and connecting rods by a three throw crankshaft having a bearing at each end, Fig. 24.8(a).

It was found that both support bearings had failed but the tribologist was able to define two important features: (a) the bearing size, its load and lubricant were such that increased speed would enhance rather than impair performance, and (b) the bearing failure initiated at the inboard edges of both bearing, indicating shaft misalignment or bowing. With the aid of a computerised model, Fig. 24.8(b), the shaft/bearing system was examined to determine shaft deflection with higher centrifugal loads due to the increased speed. The established deflection was found to increase shaft inclination within the bearing length, causing the lubricant film to break down at the ends of each bearing, making progressive failure of the bearing inevitable.

The solution was to redesign the crankshaft so that the middle throw was nominally diametrically opposite to the other two throws Fig. 24.8(c), i.e. significantly improve the dynamic balance. After the modification, the slope of the shaft within the bearing at 9000 rpm was reduced from 1/1000 to 1/100000 (at the original speed of 4000 rpm the slope was 1/50000). Subsequently, the machines were operated at even higher speeds without failure. Clearly, in this case, the problem was not tribological but a dynamics design flaw which caused bearing failure.

3. Rolling Element Bearings (Sewage Pump Motor)

A water treatment plant uses four Archimedean spiral pumps to lift raw sewage 6 m. at a rate of 3600 m^3/hr. Each pump has a closed spiral inclined at 30 degrees to the horizontal with the outlet discharging into a trough perpendicular to the pump axis. The drive reduction gearbox, is belt driven by a 100 H.P. electric motor located above the trough, Fig. 24.9(a).

The drive motor design Fig. 24.9(b), follows that of vertical motors where the rotor weight is supported on a taper roller bearing at the non-drive end, Fig. 24.9(c), with a plain radial roller bearing at the drive end to accommodate the combined redial components of rotor dead weight and drive belt tension. The design calculations indicated that both borings were well within their rated capacity with a predicted life in excess of 15 years. However, within six months, the taper roller bearing failed catastrophically with broken cages, broken rollers and indented races. The users diagnosed lubricant failure. A second bearing failed in exactly the same way within a further six months even though lubrication procedures were strictly observed and monitored. A tribology consultant was then invited to examine the problem.

The bearing debris indicated that lubrication had been perfectly adequate, and there had been no ingress of foreign matter and that failure was probably due to incorrect loading and loss of location. Wear tracks and chipped ends of rollers in the radial bearing suggested

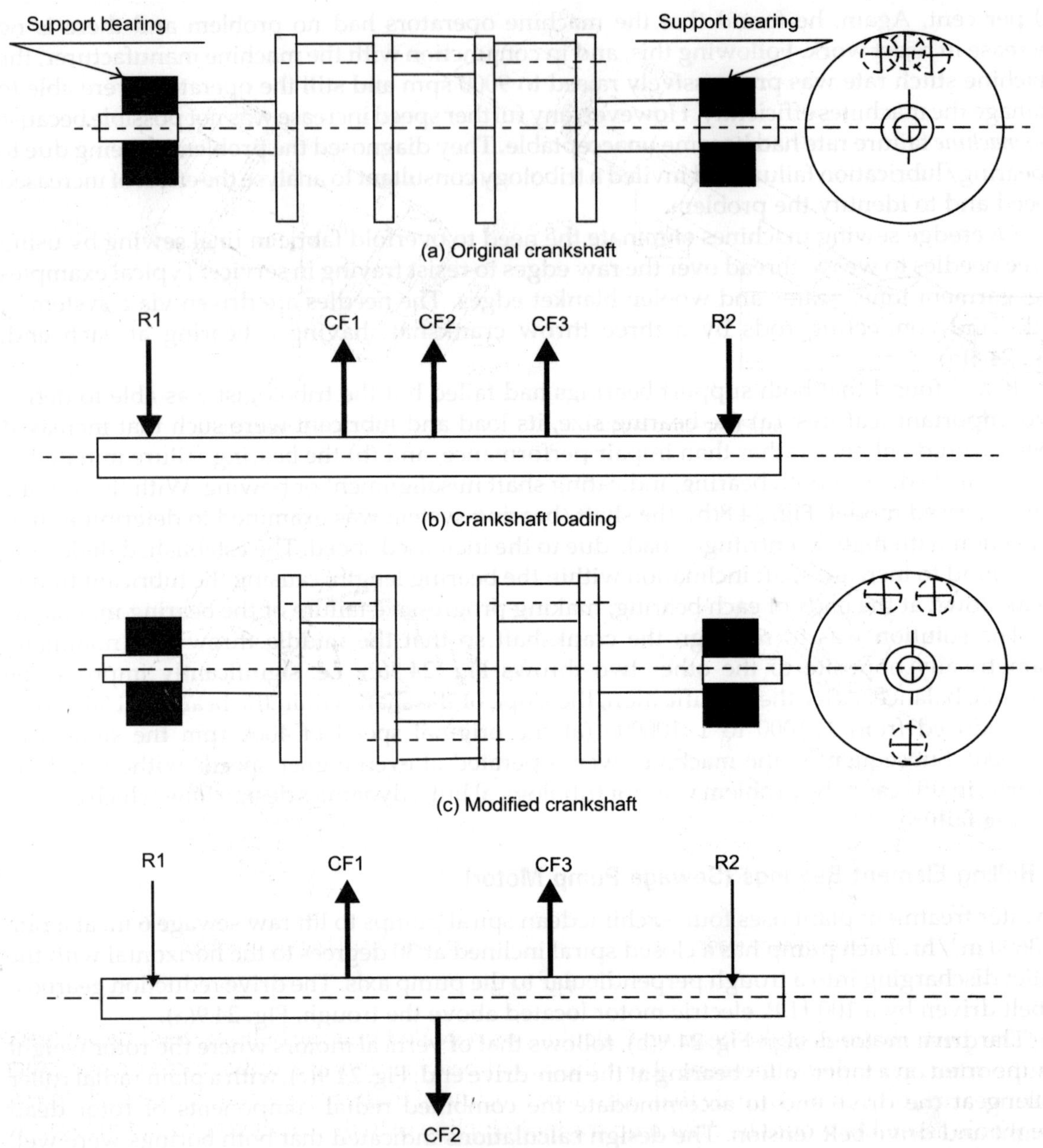

(a) Original crankshaft

(b) Crankshaft loading

(c) Modified crankshaft

(d) Reduced centrifugal loading to reduce bow

Fig. 24.8

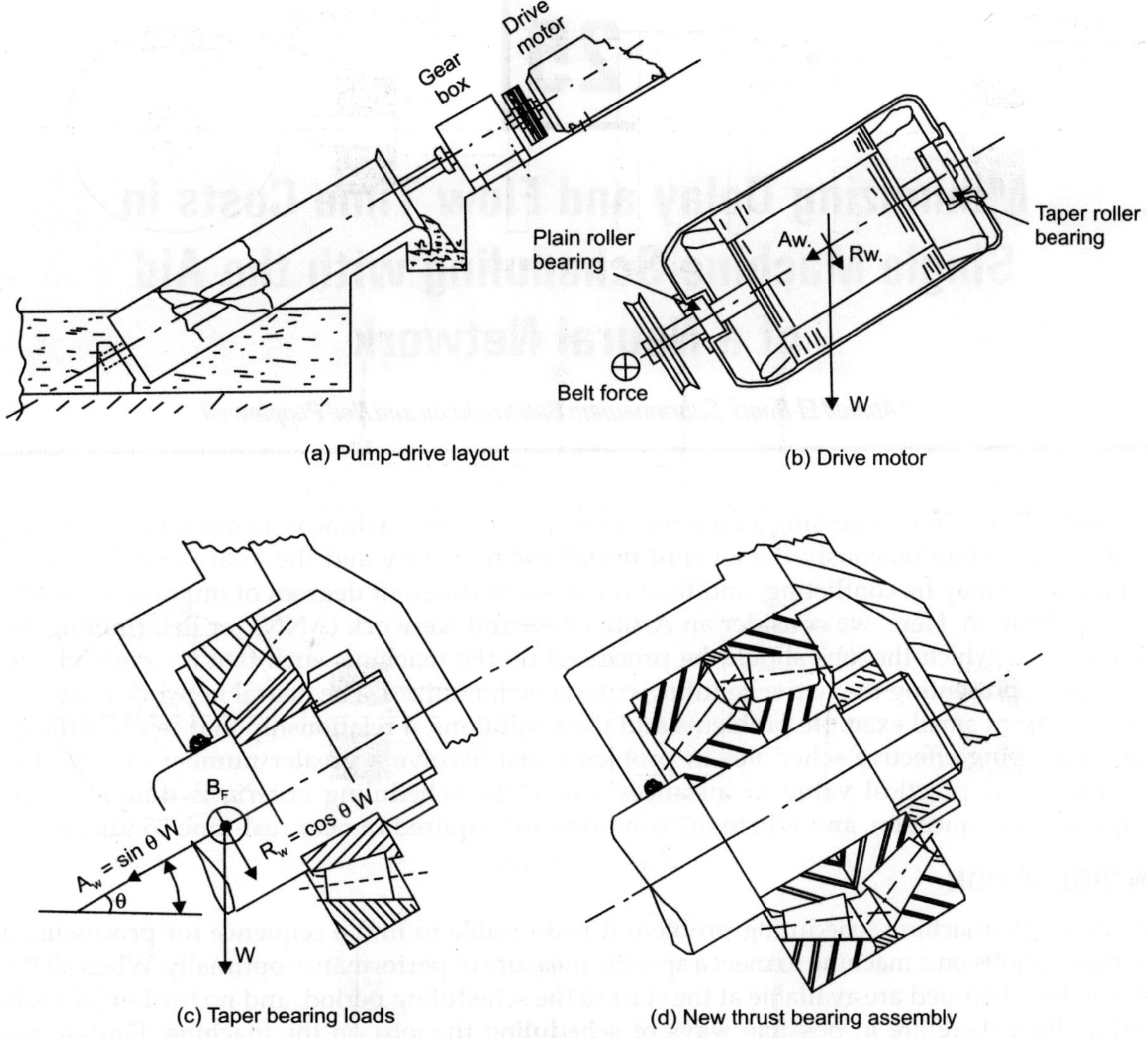

Fig. 24.9

angular perturbations of the bearing. Calculations revealed that the belt imposed radial force on the rotor, provided a moment about the radial bearing giving a radial force on the taper roller bearing of such magnitude that, if the taper incline of the bearing is considered frictionless, the whole assembly would be driven up the taper incline with the result that all radial location is lost. In the limit the bearing would "hammer" around the outer race, drop back into position and then climb out again with disastrous results. The assumption that a roller race is frictionless in the plane of a roller axis is valid, for, as a well lubricated roller rotates about its axis it can easily roll/spiral axially along the outer race unless restrained by a positive location.

The problem was solved by installing a counter thrust taper roller bearing behind the existing support bearing, Fig. 24.9(d). After modification, the motor operated for over five years. Yet again, the tribological problem can be attributed to an error in the original design where the effects of all the forces imposed on the bearings had not been fully considered, or understood.

25

Minimizing Delay and Flow Time Costs in Single Machine Scheduling with the Aid of a Neural Network

Ahmed El-Bouri, Subramaniam Balakrishnan and Neil Popplewell

A good schedule for processing a number of jobs on a single machine is, in many cases, the one that reduces simultaneously the level of in-process inventory and the total job delay. These dual criteria may be conflicting, and they can assume different degrees of importance in any one application. Here, we consider an Artificial Neural Network (ANN) for determining the sequence in which the jobs should be processed on the machine, such that a weighted cost function representing the two scheduling criteria is minimized. The neural network is trained to learn, from small example problems and their solutions, a relationship that can be utilized for developing effective schedules in problems that involve a greater number of jobs. The approach has practical value in instances where the scheduling criteria is unique to the application in question, and where the solutions are required in near real-time conditions.

INTRODUCTION

In the single machine scheduling problem it is desirable to find a sequence for processing a number of jobs on a machine to meet a specific measure of performance optimally. When all the jobs to be scheduled are available at the start of the scheduling period, and no further jobs will arrive, then there are $n!$ possible ways of scheduling the jobs on the machine. Finding the optimal schedule involves implicit or explicit evaluation of all $n!$ sequences, a task that increases in complexity as n grows. Mathematical programming techniques are available, but their efficiency deteriorates rapidly as n increases beyond 15 to 20 jobs. For problems having large n, heuristics based on adjacent pairwise interchanging (API) methods, simulated annealing and taboo search are preferred because they are not tightly restricted in applicability to specific performance criteria. Such heuristics typically start with an initial solution and implement improvement strategies.

In many real world instances, performance measures are a combination of different, often conflicting criteria. Commonly, there is a need to reduce both flow time and delay simultaneously. In this paper, an approach for solving the problem with the aid of neural networks is

*Department of Mechanical Engineering, University of Manitoba. Winnipeg, Manitoba, Canada.

demonstrated. The neural network is trained to learn a functional relationship between jobs, represented by descriptors, and the positions occupied by those jobs in the optimal sequence. A job descriptor is a vector containing statistical information regarding a job and its relation to the other jobs that it is to be scheduled with.

PROBLEM STATEMENT

Given n jobs ready at time $t = 0$ to be processed on a single machine, it is desired to find a sequence, S, that minimizes

$$Z = \sum_{t=1}^{n} [t_i T_i(S) + h_i C_i(S)] \qquad \ldots(1)$$

where d_i = Due date for job i;
$C_i(S)$ = Time job i is completed after sequence S is started at time $t = 0$;
$T_i(S)$ = Max $[0, C_i(S) - d_i]$;
h_i = Cost of holding job i for one time unit; and
t_i = Penalty per unit time for the delay of job i.

Z is a cost function that incorporates holding (time spent in system) costs and delay (amount of time past due date) costs as a function of job completion time.

The neural network that is proposed for the single machine problem is a back propagation network (BPN) having three layers of processing units. The three layers include an input, an output, and a hidden layer. The scheduling problem is represented by means of 11-triple vector descriptors for each job in the problem set. The descriptors are presented to the network at the input layer and, hence, eleven units are needed at the input layer. The input units contain the following data:

unit 1 $= p_i/M_p$ ...(2a); unit 2 $= d_i/M_d$ ...(2b); unit 3 $= SL_i/M_{SL}$ ...(2c);

unit 4 $= h_i/10.0$ ...(2d); unit 5 $= t_i/10.0$ ...(2e); unit 6 $= \bar{p}/M_p$ ...(2f);

unit 7 $= \bar{d}/M_d$ ...(2g); unit 8 $= \overline{SL}/M_{SL}$ ...(2h);

unit 9 $= \sqrt{\dfrac{\Sigma(p_i - \bar{p})^2}{n \times \bar{p}^2}}$ ...(2i); unit 10 $= \sqrt{\dfrac{\Sigma(d_i - \bar{d})^2}{n \times \bar{d}^2}}$ ...(2j); and

unit 11 $= \sqrt{\dfrac{\Sigma(SL_i - \overline{SL})^2}{n \times \overline{SL}^2}}$ ...(2k),

where p_i = Processing time required by job i on the machine;
SL_i = Slack for job i $= (d_i - p_i)$;
M_p = Longest processing time among the n jobs $= \max [p_i]\ i \in n$;
M_d = Latest due date of the n jobs $= \max [d_i]\ i \in n$; and
M_{SL} = Largest slack for the n jobs $= \max [SL_i]\ i \in n$.

The output layer consists of a single unit that assumes a value between 0.1 and 0.9. The value of the output unit is an indication of wherein the sequence the job represented at the input layer lies. Low output values indicate high priority and, hence, leading positions in the sequence. The number of units in the hidden layer is determined during the training phase.

Training the Neural Network

The steps leading to a trained neural network are the following:

(a) Generate a random set of example problems of a size n for which optimal solutions are found conveniently.

(b) Find the optimal solutions for the example problems.

(c) Select the input-output training patterns from the solved problems. The input patterns are determined from equations (2a) – (2i). The target output, G_i, for the job occupying the ith position in the optimal sequence is found by

$$G_i = 0.1 + 0.8\left(\frac{i-1}{n-1}\right), \qquad i = 1, \ldots, n \qquad \ldots(3)$$

(d) The neural network is trained by using the back propagation algorithm.

After completion of training, the neural network is ready to produce schedules. For the problem of sequencing n jobs, n job descriptors are presented, one at a time, at the input layer. For each descriptor a real value between 0.1 and 0.9 is produced at the output unit. Sequencing the jobs according to non-decreasing order of their output values produces the neural network sequence. This sequence is then subject to an adjacent pairwise interchange, based on the neural output values, to give the final sequence of jobs. In this post processing procedure, the two jobs that have the closest output values are interchanged. If the interchange contributes positively to minimizing Z, the interchange is maintained. This is repeated for the pair of the jobs having the next closest neural output values, and so on until no further interchanges improve the sequence.

Example

The neural network solution will be illustrated for the following case:

Jobs to be processed on a single machine have processing times and due dates coming from the uniform distributions U[l, 100] and U[0.4 P, 0.6 P] respectively, where P is the sum of all job processing times. This due date distribution defines a narrow range of values, indicating that the problems come from a certain segment of the population. Furthermore, the cost of delay is threefold of the holding cost per equivalent unit of time. The scheduling criterion is to determine S that minimizes

$$Z = \sum_{i=1}^{n} [3T_i(S) + C_i(S)] \qquad \ldots(4)$$

This cost function represents, for the range of processing times considered, a reasonably balanced case for the two criteria. If the holding costs were higher, for example, then sequences minimizing the mean flow time would begin to predominate, and vice versa with respect to delay. A neural network with nine hidden layers (thus a 11-9-1 architecture) is trained by using 4000 sets of input-target patterns generated randomly with integers from the uniform distributions U[1, 100] for processing time and U[0.4 P, 0.6 P] for due date. The network is satisfactorily trained after 300 training cycles. It is then tested with random data generated in a manner similar to that used for the training sets (but with different seeds). Each test set consists of 100 problems for a specific n. The neural network is evaluated by comparing its performance with the API procedure and its modified versions. The results for selected values of n are presented in Table 25.1. Table 25.1 shows the averages of Z for both the neural and API methods, as well as the number of times each method obtained the better solution. The average CPU time in each test case is also given in Table 25.1.

The neural network is able to provide, after a post-processing stage, high quality solutions to single machine scheduling problems having the objective of minimizing a cost function combining delay and flow time criteria. An example problem demonstrated that the neural network was quicker and better than a comparable method that implements a more exhaustive series of API strategics. In conclusion, the neural network described here is a competitive model with application potential for problems having a specific character and performance criteria.

Table 25.1. A comparison of the neural network results with API

n	*Neural network (Z)*	*API (Z)*	*Best solution neural*	*Best solution API*	*Neural CPU time (s)*	*API CPU time (s)*
5	811.85	809.46	91	99	0.001	0.004
10	2457.64	2465.81	78	79	0.005	0.013
20	8400.02	8474.34	67	51	0.015	0.098
30	17835.46	18122.17	66	35	0.037	0.436
40	31245.98	31993.32	83	17	0.109	1.345
50	48142.34	49113.34	82	18	0.257	3.196
60	67643.67	68937.42	79	21	0.523	6.707
70	91555.24	93882.52	90	10	1.102	12.530
80	119777.84	122608.35	93	7	1.931	20.408
90	148507.43	152120.24	89	11	3.296	31.794
100	183491.00	188169.80	92	8	5.120	54.003

26

Providing Hands on Experience in the Area of CAM (Computer Aided Manufacturing) in University Level Courses

*Dr. S. Balakrishnan**

The application of Computer Integrated Manufacturing (CIM), Robotics and Automation and its increased adoption in a number of industries has necessitated universities to introduce these fields as part of their curriculum. Many universities have responded to these needs by introducing new courses in their curriculum. However, unlike many traditional fields of engineering, these areas can benefit by providing intense hands on experience in the laboratory session that accompany the course. Can this be accomplished with limited finding? This paper will discuss the experiences gained in establishing such a facility at the University of Manitoba, Winnipeg, Canada. The presentation will also provide details of a case study from industry that was analyzed as part of the undergraduate laboratory instructions.

INTRODUCTION

Computer Aided Manufacturing (CAM) and Computer Aided Design (CAD) involves the use of digital computers to automate a wide variety of function in design and production. CAD is concerned with the implementation of computers for designing a part and CAM is concerned with the use of computer to support manufacturing engineering activities. CIM includes all of the engineering functions of CAD/CAM. The barrier and distinction between these two entities is constantly changing. Universities cannot really keep up with all of the changes. However, something must be done to still prepare students to understand this high level technology and it must be done quickly and in the most efficient way. With the advancements in digital technology and automation more and more industries are adopting these technologies to stay competitive in the global market. One should keep in mind that the primary objective of engineering institutions is to train engineers and not operators. The students must be able to grasp the principles and be given the opportunity to think and apply the fundamental principles to solve complex problems. The author of this paper will discuss one area of CIM, namely automated manufacturing and how it has been possible to provide hands on experience with limited resources in a university environment with a few examples.

**Professor and Head, Department of Mechanical and Industrial Engineering, University of Manitoba, Winnipeg, Manitoba, Canada.*

Structure of CIM Programme

The Department of Mechanical and Industrial Engineering at University of Manitoba, Winnipeg, Canada, offers two-degree granting programmes, namely Mechanical Engineering and Industrial Engineering. The first two years are common to both programmes and specialization starts from the third year of the four-year programme. The author offers two undergraduate courses at the final year of Industrial programme that cover specialized topics in the area of CIM. The first course is titled "Computer Aided Manufacturing and Robotics". It covers topics such as Computer Numerical Control (CNC), application of computer aided design software for CNC applications, programmable logic controllers (PLC) and process automation applications, and robotics, the second course, namely "Computer Integrated Manufacturing and Automation" built upon the foundation given in the first course and covers a wide variety of topics such as: group technology, manufacturing resource planning, geometric modeling, material handling devices, cellular manufacturing, flexible manufacturing systems and other related topics. Both courses are core courses for Industrial Engineering programme and are optional for Mechanical Engineering programme. The average enrolment is thirty-five and twenty-five students respectively. The existing facility does not permit greater enrollment. Both courses have 3 hours of lecture and 3 hours of laboratory session per week. The students are divided into groups of five and several sessions are conducted throughout the week. Although it demands lot of work hours, it has facilitated more individual hands on experience for the students.

Existing Facility

The facility is located in the Engineering Faculty and has two industrial five axes robots, four table top robots, one industrial CNC machine and five table top CNC machines, ten work stations with CNC and CAD software, several PLCs, a wide variety of fixtures, and sensors. It should be noted that all these machines are primarily geared for stand-alone operation. Some of the equipment was received as donation from local industries and government agencies. The laboratory also supports a number of research projects.

Organization of Laboratory Sessions for Stand Alone Machines

As mentioned earlier, both courses have emphasis on hands on experience in the laboratory sessions. In the first course, the students deal with individual components of machines while in the following course, integrated operation is focussed. This difference in emphasis has brought a real challenge, which will be the main focus of this paper.

Laboratory Related to CNC Machining

The students work with several modules during the laboratory sessions in the first course. They start off familiarizing with aspects related to computer aided programming for CNC machines. The software currently used is "MasterCam" which runs on a PC environment. It is capable of creating CNC codes for multi-axis machining. Given the fact that the duration of the course is only thirteen weeks, a variety of problems exposing the students to two and half-dimensional as well as three-dimensional machining is given first. The students are provided with a manual that allows them to work through several sample problems that demonstrate contour, pocket and surface milling of varying degrees of difficulty. The students are encouraged to understate clearly the example problems. Towards the end of the semester, the students are given part drawing of two more parts from the real world. The part to be machined will include all aspects of milling. The students are given minimal guidance. It is their responsibility to measure key dimensions of the part, create the geometry of the part on the CAD system, and decide on machining sequence as well selection of tool path. This has proved to be real design challenge in machining and use of CAD for CAM.

At every stage of the design process, the students are encouraged to verify the results using built in simulation modules. When they are ready to machine, a three axes CNC mill is used to cut the part and analyze the results. The challenge has been to identify parts that pose a real challenge to the CAD as well to the students. The solution must be obtained within the given time. They should also incorporate as many of the milling operations and produce a part with good quality with minimal tool change and near optimal machining.

Programmable Logic Controller

The emphasis of this laboratory session is to first get familiar with PLC, solve simple problems and then proceed to solve more complex problems. Two models were built for hands on experience. The first model demonstrates use of PLC for ventilation control while the second demonstrates the application in an automated drilling operation. Both the models were built using parts salvaged from used equipment. Use of appropriate sensors for either a process control or automating a process is also demonstrated. Both the problems are real world applications and the models mimic their operation. The students are required to write and compress Boolean logic equations, implement and test their solutions. The solution requires extensive use of timers, latches, Boolean logic operators and counters. The problem has many solutions and the students come up with a wide variety of solutions.

Robotics

The focus is to provide the students with hands on experience in programming a five-degree of freedom industrial robot. Several stand-alone tasks are presented that allow the students to programme, understand the features, limitations and efficient use of the various degrees of freedom within the work envelope of a robot. The knowledge gained is effectively used in the more difficult task that follows in the second course.

To summarize, the scope of the laboratory sessions in the first course is primarily to provide hands on experience on stand-alone equipment that form part of a Computer Integrated Manufacturing cell. The challenge has been on finding appropriate examples for providing hands on experience that ties what is covered in the theory classes. What is not focussed in the first course is an equally important aspect of automation, namely integrated operation of several computer controlled machines, jigs and fixtures and their role in automating a process. This is brought out in the second course, which is described in the next section.

Organization of Laboratory Sessions for Integrated Operation (CIM)

As mentioned earlier, the facility has several machines with stand-alone controllers. It took the author a number of years to integrate several stand-alone controllers for a total computer integrated operation. Integration has been achieved using PC and PLC as cell controllers. A wide variety of sensors are utilized for monitoring the status of machines which then allow coordinated operation using master cell controller. Having achieved this, the challenge is to produce a product that requires several operations within the cell. The task described below is one of several model problems that have been put through the cell.

At the beginning of the course, a process from a local industry (that has not been automated) is chosen. For the past two years the problem described below has been focussed.

Problem Definition

A local industry is involved in the manufacture, fabrication and finishing a wide variety of hydraulic cylinders for use in front-end loaders and miscellaneous farm machinery. Altogether fifty-seven different types of cylinders varying in size are manufactured. The problem focuses on the last stage of the operation and, in particular, the heaviest cylinder. The cylinder focussed weighs 62 kg, approximately 2 meters long and about 10 cms in diameter. The operation starts with the cylinders being transported to the final location in batches of 20 to 25, on a wooden crate. They are normally stacked in 2 to 3 layers and there has been no attempt to load them onto the crate with no fixed orientation. The cylinders have clevice on one side and hook on the other side. The operation then involves two people loading these cylinders one by one onto a moving conveyor. The conveyor is in continuous motion at a speed of 80 cm/min. The conveyor has 210 hooks set 45 cms apart and will arrive at the loading site in no fixed orientation. The operation of lifting, finding the orientation of the hook and then hanging the cylinders is all done manually and has been a tedious job. The workers involved have been experiencing severe back problems from time to time. This has resulted in lost workers time, as well as monetary compensations. When the line is fully loaded, each cylinder is rinsed, washed with a special detergent and left on the line to dry. This is followed by manually painting each cylinder at a fixed location on the line. This process is also done with the conveyor in motion. The process concludes with manually unloading all the cylinders when the paint is fully dry.

The following question is posed to the students. Can this operation be automated and, if so, at what cost? Fully analyse this process and provide a solution. The students are given roughly 15 hours to brainstorm in-groups of five. At the conclusion of this exercise they are required to write a report suggesting a solution to this problem from a point of automating the process. The report should include a brief write up on the proposed solution together with sketches on various components required to implement the solution. Use of sensors (contact as well as non-contact) is encouraged to implement unmanned operation of the line. The students have to build all the jigs and fixtures proposed in their design within the limited time.

The students are provided with wooden dowels with clevice on one end to mimic a cylinder. At the end of 15 hours, the solutions proposed by various groups are examined by the instructor and teaching assistants to track any serious problem with the solution. In the remaining 18 hours, the students are required to (in individual groups) test their methodology and conclude with a written report on the success and deficiencies of the proposed solution.

A Sample Solution

A methodology proposed utilized a robot, a positioning device with sensors to detect when the cylinder is in correct orientation for pick up by the robot, a small scale conveyor, custom designed hooks, custom designed fixtures for correctly orienting the hooks, sensors to make sure that hooks are in correct orientation, and devices to hold the cylinders in place before actual loading of the cylinders. A layout of the system is shown in Fig. 26.1.

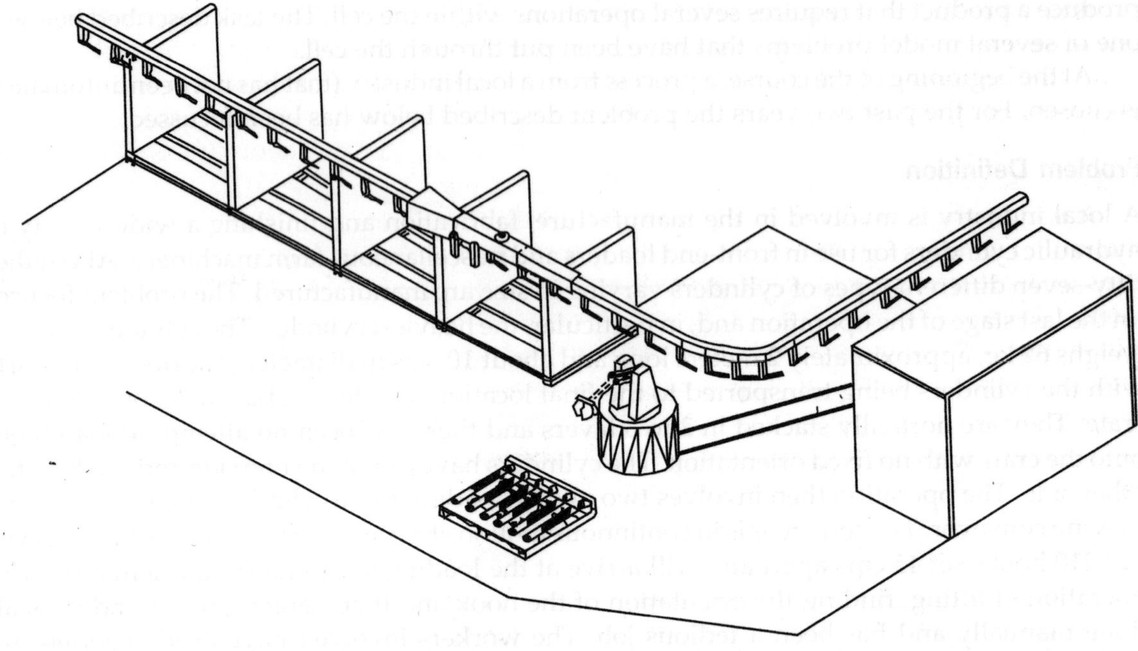

Fig. 26.1. Layout of cell

The cylinders arrive at the loading station. The cylinders are in all possible orientation. Unlike human, who can use his visual sensory information, the robot is not equipped with vision capability. Hence, the task is to pick up the cylinders from the pallet and position them in a device that will bring them to the correct orientation. This device is shown in Fig. 26.2. The positioning device has a motor, which is in continuous motion with a clutch to engage and disengage whenever the motor needs to drive a drive train. The drive train powers two sets of rollers, which rotate the cylinder whenever a cylinder is positioned on the rollers. It is also necessary to receive a message from the robot for this process to be initiated. The clutch will disengage the drive mechanism only when the cylinder reaches a desired orientation. This was detected using a set of non-contact sensors, which get activated whenever the cylinders reach the correct orientation. The positioning device was made from components retrieved from an old printer. The process depicted so far required a PLC to automate and control this stage of the process and the students are required to design the PLC logic, implement and test.

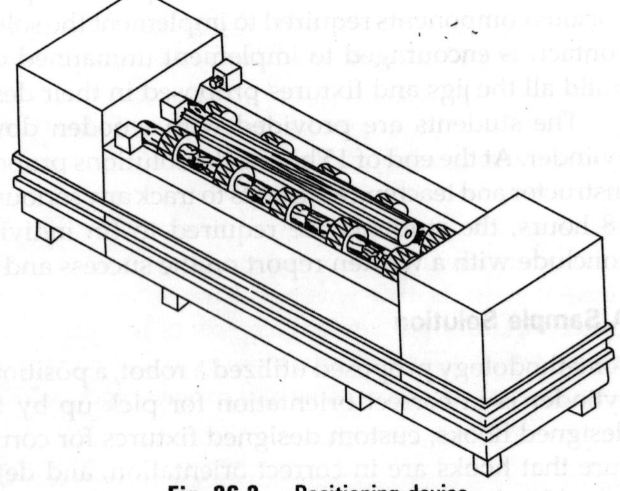

Fig. 26.2. Positioning device

When the PLC determines that the cylinder is in the desired orientation, it sends a message to the robot to proceed to the next stage. The robot then picks up the correctly oriented cylinder using suction powered grippers and moves to the vicinity of the moving conveyor. It waits for a hook to arrive at a desired orientation. One solution will attempt to load the hook directly. An alternate solution leaves the cylinder in a temporary fixture that will allow automated loading of the hook whenever a cylinder is present in the fixture.

A sketch of one of the specially designed hooks is shown in Fig. 26.3. The hooks were made from paper clips and are attached to the conveyor chain using standard screws and allow free rotation. Devices that will permit consistent reorientation of all the hooks to a certain desired orientation were part of the design requirement. The students are also required to programme the robot to accomplish the various tasks, such as loading the positioning device, waiting to unload, load the line with the cylinder in the right orientation and unload when the cylinders are ready. Although the task appears to be easy, the students discover many limitations posed by the work envelope of the robot and end up modifying their design.

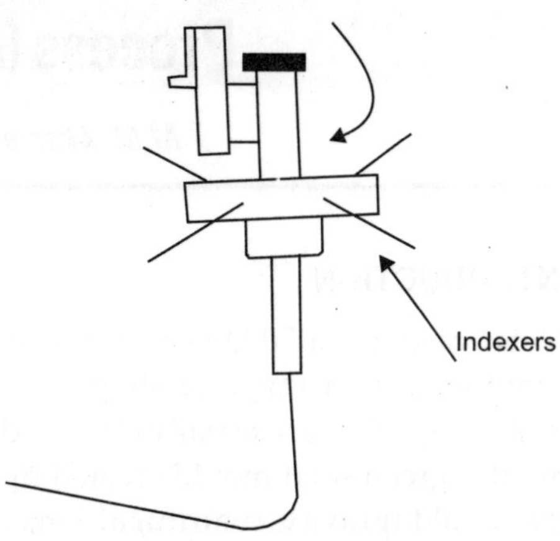

Fig. 26.3. Proposed hook

All the fixtures were built out of plywood, foam, miscellaneous shop materials and are held together using hot glue, wood glue and duct tape. Some of the mechanisms used were salvaged from discarded machinery donated by local industry.

The final phase involved integration of the various aspects of the design followed by proposed testing of the methodology. With all the components in place, robot programmes and control programmes were fine-tuned and the testing was done.

No group succeeded in getting a perfect working model. Many of the ideas proposed sounded workable needed modifications. They were amazed how from a conceptual to the realization of a design was a big step. It was the general agreement that in the conceptual design one is not able to foresee clearly all the problems that contribute to a successful implementation. There was also a big difference in making a design work with a high level of reliability versus getting it to work sometime. Making small-scale models of devices was not easy. Maintaining high tolerances in fabrication stages was also found to be a major factor in the overall success. Small process variability also contributed to failures. It was also agreed that revisions could be made in the design to account for these variations in the process parameters and evolve a design, which will compensate for minor variations in them. Majority overlooked this aspect during the initial design iteration. It was the general opinion that working on a project, such as this was much more educational than merely stopping after proposing a conceptual design. Several aspects of integration that could not be taught in a lecture-based presentation were brought to light in this type of exercise. To summarize, as an instructor the results were very rewarding and provide additional motivation to pursue this type of instruction.

27

Casting Process Selection Using Analytic Hierarchy Process (AHP) and Fuzzy Logic

M.M. Akarte *, B. Ravi ** and Robert C. Creese ***

INTRODUCTION

Metal casting—a 7000 year-old technology—offers the widest variety of routes to produce component in a range of shapes, sizes, metals, quantities and quality requirements. These routes (Fig. 27.1) are usually classified in terms of the material and the method of producing the mould (green sand mould created by a permanent pattern, etc.) and how the metal flows into the mould (gravity, centrifugal force, vacuum, low pressure, high pressure).

Each casting process has its own capabilities, advantages and limitations in terms of design, quality, production and cost characteristics. For example, a particular process may not be able to produce a part with wall thickness below a critical value, The critical wall thickness for Aluminum parts is 3.0 mm by sand casting process and 0.75 mm by pressure die casting process. The corresponding values for steel parts are 4.75 mm by sand casting and 0 mm (cannot be produced) by die casting. Evidently, the dependence of process characteristics on the type of cast metal and often even on the class of application leads to a large number of sets of process characteristics. In practice, the process characteristics also depend on the equipment, manpower skills, quality management practices, and other company-dependent factors. For example, minimum section thickness of copper-based parts that can be produced by permanent mould casting can vary from 2.5 mm to 8.0 mm. Most of the characteristics are therefore indicated as a band of values (minimum, maximum and most common values) as available in technical literature. Further, some process characteristics are described qualitatively (for example, surface detail capability of investment casting process is HIGH), adding to the difficulty of representing all characteristics in a uniform manner.

Selecting the most suitable process for the given product requirements is a critical step in product life cycle. This influences all downstream decisions, such as the type of tooling, vendor selection, machining operations and quality control procedures. These, in turn, affect the

* Production Engineering Department, SGGS College of Engineering & Technology, Nanded, India.

** Department of Mechanical Engineering, Indian Institute of Technology, Mumbai, India.

*** Industrial & Management Systems Engineering, West Virginia University, Morgantown, WV, USA.

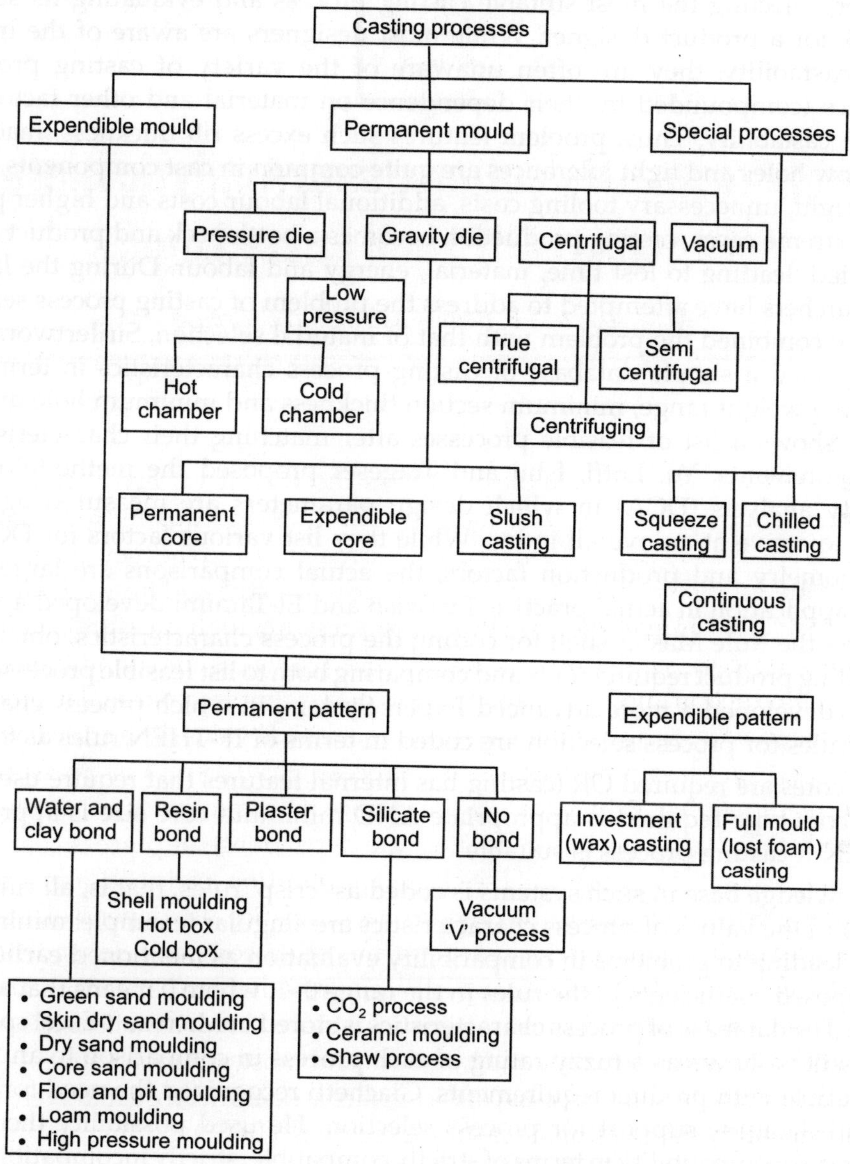

Fig. 27.1. Types of casting processes

tooling and labour costs, quality assurance and total lead time; these final factors together are referred to as 'castability'. An early and informed decision about the casting process will not only ensure the castability of a given design, but also provide an opportunity to improve its castability through suitable design changes. In other words, product design and process design must proceed concurrently to achieve overall optimization in the shortest time. Indeed, Design for Manufacturability is well established in most engineering sectors, particularly for machined and plastic components.

However, selecting the most suitable casting process and evaluating its suitability is a difficult task for a product designer. While most designers are aware of the importance of design for castability, they are often unaware of the variety of casting processes, their characteristics (compounded by their dependence on material and other factors) and their influence on castability. Thus, problem features such excess rib thickness, inadequate fillet radius, narrow holes and tight tolerances are quite common in cast components, which result in excess weight, unnecessary tooling costs, additional labour costs and higher percentage of defects. In extreme cases, casting production becomes a bottleneck and product designs have to be modified, leading to lost time, material, energy and labour. During the last ten years, several researchers have attempted to address the problem of casting process selection. Some of these have combined the problem with that of material selection. Sirilertworakul, Webster and Dean created a simple database of casting process characteristics in terms of suitable metals, casting weight range, minimum section thickness and minimum hole diameter. Their programme shows a list of feasible processes after matching their characteristics with the product requirements. Yu, Lotfi, Ishii and Trageser proposed the methodology of design compatibility analysis (DCA) in which design parameters are measured against process capabilities to arrive at an overall index. While they list various factors for DCA, including material, geometry and production factors, the actual comparisons are largely subjective, limiting its application in actual practice. Darwish and El-Tamimi developed a simple Expert System using the Rule Master shell for coding the process characteristics, obtaining the user input regarding product requirements and comparing both to list feasible processes. Er, Sweeny and Kondic developed a more advanced Expert System, in which process characteristics as well as the rules for process selection are coded in terms of IF-THEN rules as follows:

IF (no cores are required OR (casting has internal features that require use of cores AND core type required is appropriate AND minimum core size is appropriate))
...THEN X casting process is suitable.

The knowledge base in such systems is coded as 'crisp' rules, that is, all rules are equally important and the values of process characteristics are singular (example: minimum core size is 6.5 mm), leading to problems in compatibility evaluation as mentioned earlier. Perzyk and Meftah proposed coefficients to the rules in the range 0–1, where 0 means that a rule does not apply at all. The database of process characteristics is stored in Microsoft Excel spreadsheet and the final result is shown as a fuzzy rating of each process in comparison to an 'ideal' process fully compatible with product requirements. Giachetti recognized the need for both decision support and database support for process selection. He used possibility theory to classify product-process compatibility in terms of strictly compatible, strictly incompatible and partially compatible, giving fuzzy values. But he used linguistic terms (very important, etc.) to assign weights to the criteria, which may lead to incorrect weights given the number of criteria involved. Lovatt and Shercliff proposed a comprehensive methodology for process selection based on initial screening, primary assessment, performance assessment and economic evaluation. While the first two steps are based on attribute matches and incompatibility checks, the next two depend on process and activity-based cost models to select the best process. They further extended their approach by using fuzzy logic to represent process characteristics and demonstrated it for process selection for aluminum parts.

This investigation addresses some of the limitations of earlier work in casting process selection and attempts to link it with a CAD system to provide an integrated environment to a design engineer so that he can, not only evaluate product-process compatibility but also use it for design improvements. In particular, this work uses fuzzy logic to model the process characteristics, and introduces the technique of Analytical Hierarchy Process (AHP)—a decision-aiding tool developed by Satty for complex and multi-attribute problems—to objectively assign the relative weights to process evaluation criteria.

CASTING PROCESS SELECTION CRITERIA

The schematic representation of the overall methodology for casting process selection is shown in Fig. 27.2. The process selection criteria are grouped, useful for systematically assigning and verifying the relative weights to the criteria using AHP. Candidate processes taken from the database are passed through the critical criteria to shortlist the feasible processes. The characteristics of feasible processes are then evaluated against the product requirements using a fuzzy logic model for objective criteria and a linear weight model for subjective criteria. The

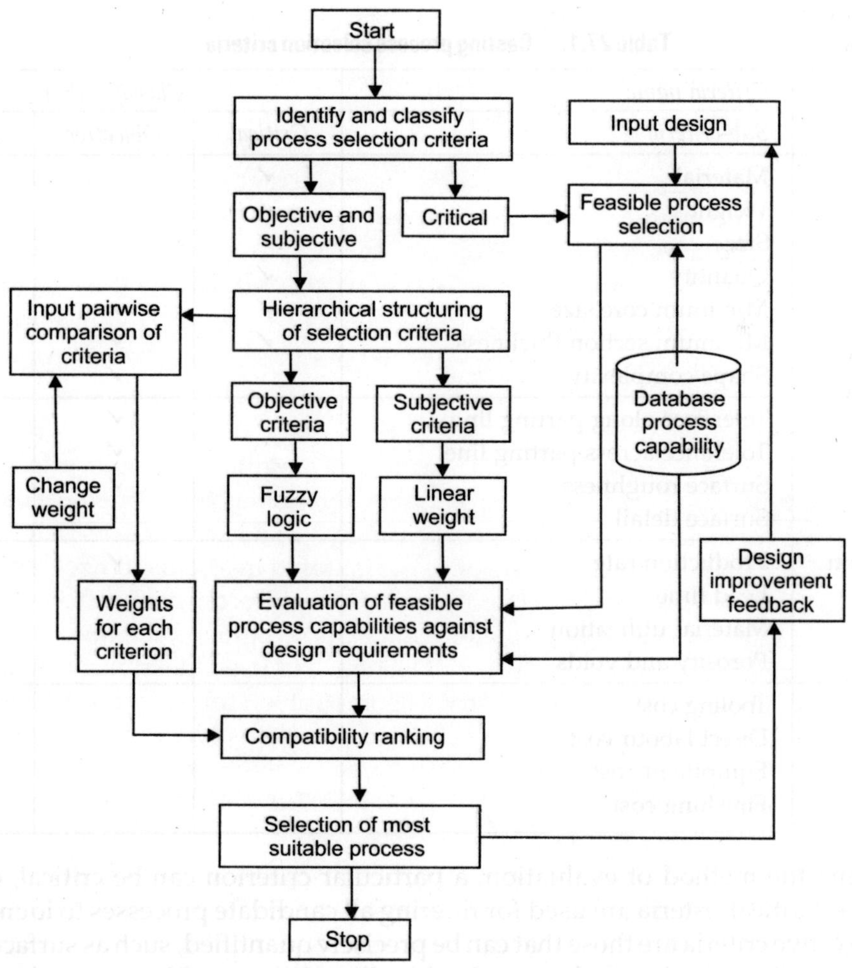

Fig. 27.2. Methodology for casting process selection

criteria weights and process evaluations together provide a compatibility index, which is used for ranking the feasible processes. After selecting a particular process, the product-process compatibility table provides pointers to design improvement for the designer.

Criteria Groups

Various criteria for evaluating the overall compatibility of casting processes with the design specifications have been identified after a detailed review of literature. A total of 19 criteria have been identified and segregated in four groups: Design, Quality, Production and Cost (Table 27.1). The design group comprises of 7 criteria: material, weight, size, quantity, minimum core size, minimum section thickness and shape complexity. The quality group comprises of 4 criteria: tolerance along and across the parting line, surface roughness and surface detail. The production group comprises of 4 criteria: production rate, lead time, material utilization and porosity-voids. The cost group comprises of 4 criteria: tooling cost, direct labour cost, equipment cost and finishing cost.

Table 27.1. Casting process selection criteria

	Criteria name	*Classification*		
Group	*Sub-criteria*	*Critical*	*Objective*	*Subjective*
(1) Design	Material	✓		
	Weight	✓		
	Size	✓		
	Quantity	✓		
	Minimum core size	✓	✓	
	Minimum section thickness	✓	✓	
	Shape complexity		✓	
(2) Quality	Tolerance along parting line		✓	
	Tolerance across parting line		✓	
	Surface roughness		✓	
	Surface detail			✓
(3) Production	Production rate		✓	
	Lead time			✓
	Material utilization			✓
	Porosity and voids			✓
(4) Cost	Tooling cost			✓
	Direct labour cost			✓
	Equipment cost			✓
	Finishing cost			✓

Considering the method of evaluation, a particular criterion can be critical, objective or subjective type. Critical criteria are used for filtering all candidate processes to identify feasible processes. Objective criteria are those that can be precisely quantified, such as surface roughness in microns. Subjective criteria are characterized by linguistic variables (example: tooling cost HIGH).

Estimation of Weights

The relative weights to casting process selection criteria are determined using AHP. This involves pairwise comparison of only two criteria at a time and assigning the importance (Table 27.2) to the first with respect to second criteria. As a result, the importance of ith criterion over j^{th} criterion is given by a_{ij}, and importance of jth criterion over i^{th} is given by $a_{ji} = 1/a_{ij}$. The steps to find the vector of weights and check the consistency of judgement, as suggested by Satty, are included in the Appendix.

Table 27.2. Scale of relative importance

Relative Importance (a_{ij})	*Definition*
1	Equal importance of i and j
3	Weak importance of i over j
5	Strong importance of i over j
7	Very strong importance of i over j
9	Absolute importance of i over j
2, 4, 6, 8	Intermediate values

The AHP approach is used to assign weights to the group criteria (C_1, C_2, C_3, C_4) as well as weights to individual criteria in a particular group C_{jk} ($j = 1, 2, 3, 4; k = 1, 2, ... SC_j$; where, SC_j is the total number of criteria belonging to j^{th} group). The sum of weights of individual criteria in a particular group is normalized to one. The sum of weights of the four criteria groups is also normalized to one. Thus, the real weight of any particular criterion is equal to the product of its own weight and the weight of the criteria group.

PRODUCT-PROCESS COMPATIBILITY

Product requirements include specifications related to its geometry, material, quality and quantity. These are obtained interactively from the user or by interfacing with a casting design system as described later in this paper. For a given product requirements, feasible casting processes are short-listed by filtering all candidate processes available in the database using the critical criteria. Then the compatibility ranking of feasible processes (p_1, p_2,..., p_n) for a given product is obtained as follows:

(1) Determining the compatibility value of feasible process characteristics with respect to each criterion,
(2) Obtaining the compatibility index of a feasible process, given by the sum of the products of compatibility values and the respective weights.
(3) Ranking feasible processes by sorting them according to their compatibility index.
(4) Suggesting compatibility improvement through design changes.

Evaluating Process Capability

As described earlier, process evaluation criteria are of two types: objective and subjective. These are evaluated by using a fuzzy logic model and a linear weight model respectively. The linear weight model is described first.

Process characteristics, which have to be evaluated against subjective criteria, are generally expressed in terms of linguistic variables (example, low tooling cost, very high finishing cost or lead-time is days-to-weeks). Quantification of these qualitative rating is mapped to a number between [0,1]. In this work, the variables low, low to medium, medium, medium to high, high, high to very high and very high refer to the ratings 0, 0.167, 0.333, 0.50, 0.667, 0.83 and 1.0, respectively. Similarly, lead-time variables (months, weeks to months, weeks, days to weeks and days) are also mapped between [0,1].

Process characteristics, which have to be evaluated against objective criteria are not precisely defined in casting domain. Therefore, a fuzzy logic methodology is used to find out the degree of compatibility between the design requirement and imprecise values of process capabilities, which are generally defined in terms of a range. The range of process capability values need to be mapped on the common scale of [0,1] such that every value in this range will represent a compatibility value between 0 and 1 depending on the difficulty of manufacturing. Thus a process capability value corresponding to 1 will represent complete compatibility (easy producibility of process characteristic for the given design attribute). In this work, three values, viz. minimum, maximum and mean or desirable, have been used to represent imprecise process capability range into triangular and trapezoidal fuzzy membership functions (Fig. 27.3). A membership function defined on the fuzzy set Z is given by

$$Z : X \rightarrow (0,1)$$

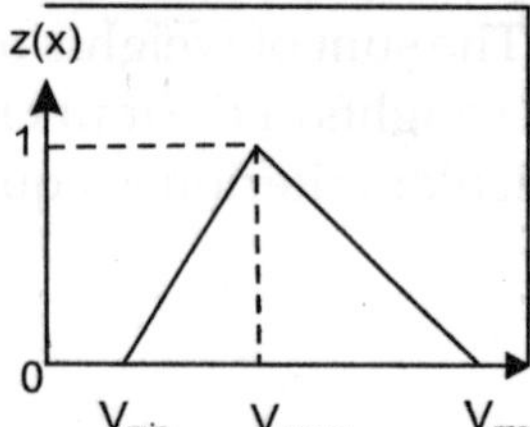

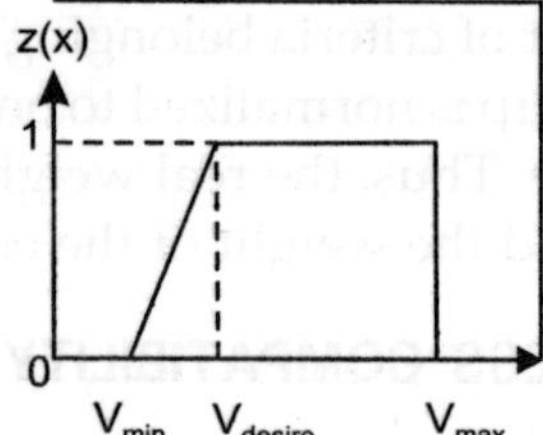

Fig. 27.3. Fuzzy membership functions: triangular and trapezoidal

where, X = A universal set of process characteristics values,
Z = Fuzzy set of all values in the defined range of process capability.

In other words, the membership function defined on fuzzy set Z will map the process capability values into a real number between [0,1]. Therefore, given the value of design attribute as x units, the process compatibility for this value using triangular fuzzy membership function Z(x) can be calculated as

$$Z(x) = \begin{cases} 1 & \text{if } x = V_{mean} \\ (x - V_{min.})/(V_{mean} - V_{min.}) & \text{if } V_{min.} < x < V_{mean} \\ (V_{max.} - x)/(V_{max.} - V_{mean}) & \text{if } V_{mean} < x < V_{max.} \\ 0 & \text{otherwise} \end{cases}$$

Similarly, degree of process compatibility for trapezoidal fuzzy membership function Z(x) can be obtained by

$$Z(x) = \begin{cases} 1 & \text{if } V_{max.} > x > V_{desire} \\ (x - V_{min.})/(V_{desire} - V_{min.}) & \text{if } V_{desire} > x > V_{min.} \\ 0 & \text{otherwise} \end{cases}$$

Graphical representations of fuzzy membership functions demonstrating capabilities of sample processes considered in this work are shown in Fig. 27.4. For example, if a particular product requires a surface roughness of 2.0 µm, then the compatibility values for sand mould, shell mould, permanent mould, die casting and investment casting processes will be 0, 0.98, 0.97, 0.52, and 0.94 respectively, indicating that the shell moulding process is most likely to produce the target surface roughness.

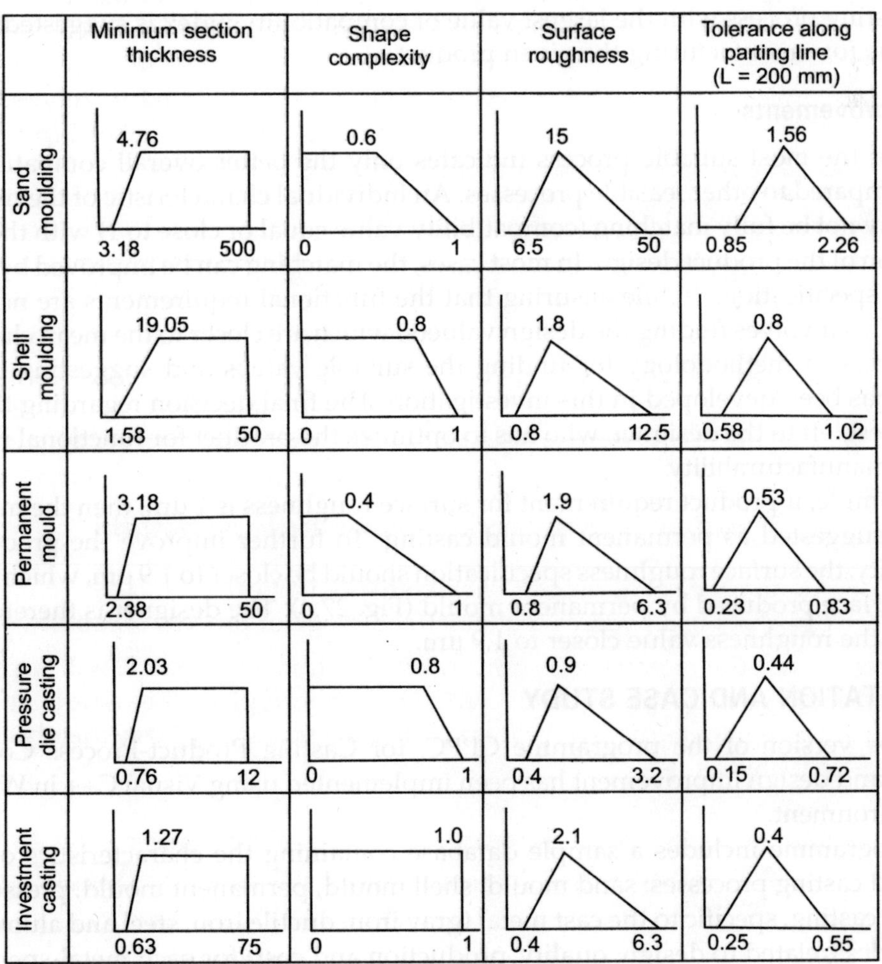

Fig. 27.4. Fuzzy membership functions for process capability (Aluminium)

Compatibility Index and Ranking

After calculating the compatibility value of each characteristic of all feasible processes for the given product design, an overall compatibility index for each process is computed as follows:

$$PC_i = \sum_{j=1}^{4} \sum_{k=1}^{SC_j} C_j C_{jk} W_{jki}$$

where, PC_i = Process capability score of ith candidate process.
SC_j = Total number of criteria belonging to j^{th} group criteria.
C_j = Importance (weight) of jth group criteria.
C_{jk} = Importance of kth criteria belonging to j^{th} group.
W_{jki} = Compatibility of ith process for k^{th} criteria of jth group.

The casting process with the largest value of compatibility index is suggested as the most suitable one for manufacturing the given product.

Design Improvements

Selection of the most suitable process indicates only the better overall compatibility of the process compared to other feasible processes. An individual characteristic of the most suitable process may not be fully matching (compatibility value equal or close to 1) with the respective specification of the product design. In most cases, the matching can be improved by fine tuning the design specifications (while ensuring that the functional requirements are not adversely affected). This involves finding the design value(s) which are closer to the mean characteristics of the process. A methodology for finding the suitable values and suggesting them to the designers has been developed in this investigation. The final decision regarding the values is however, best left to the designer, who has to optimize the product for functional requirement as well as manufacturability.

For example, if product requirement for surface roughness is 1 μm, then the most suitable process is suggested as permanent mould casting. To further improve the product-process compatibility, the surface roughness specification should be closer to 1.9 μm, which is the mean value of surface produced by permanent mould (Fig. 27.4). The designer is therefore advised to increase the roughness value closer to 1.9 μm.

IMPLEMENTATION AND CASE STUDY

A prototype version of the programme CPPC, for Casting Product-Process Compatibility evaluation and design improvement has been implemented using Visual C++ in Windows NT 4.0/98 environment.

The programme includes a sample database containing the characteristics of five most widely used casting processes: sand mould, shell mould, permanent mould, pressure die and investment casting, specific to the cast metal (gray iron, ductile iron, steel and aluminum). The characteristics (related to design, quality, production and cost) for each metal-specific casting process (example: sand_casting_aluminum) are stored in separate files. The database is independent of the programme and can be easily customized to any particular foundry. The programme has also been interfaced to a casting design system AutoCAST for importing product requirements (including geometric data such as weight and minimum thickness). The results of a case study on 'end-shield', a medium complex gray iron casting used in electric motors (Fig. 27.5) are presented here.

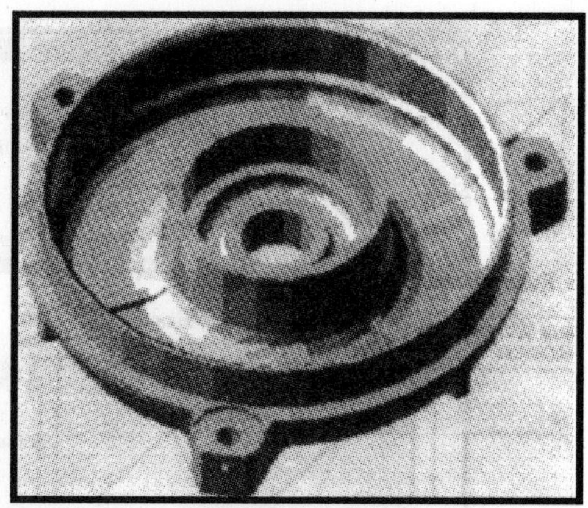

Fig. 27.5. Case study. Component: end-shield casting

After starting the programme, the user first provides (or imports) the specifications and requirements of the product (Fig. 27.6).

The next step involves pair-wise comparison of the selection criteria (first at group level and then at individual level) using AHP methodology to obtain their relative weights (Fig. 27.7). The weights, which are usually the same for a particular class of application, can be stored and later retrieved for similar applications. For the test case, the criterion for minimum section thickness has received the largest weight ($0.39 \times 0.5 = 19.5$ per cent) followed by tolerance along parting line (11.61 per cent), core hole size (9.75 per cent) and shape complexity (9.75 per cent).

Based on the input of product requirements and criteria weights, the programme short-listed sand mould, shell mould and investment casting processes. This is based on critical criteria (material—gray iron and the minimum section thickness - 3.5 mm) which eliminated other processes. The three feasible processes were further evaluated using objective and subjective type criteria to determine their overall compatibility index (Fig. 27.8). Sand casting was indicated as the most suitable process.

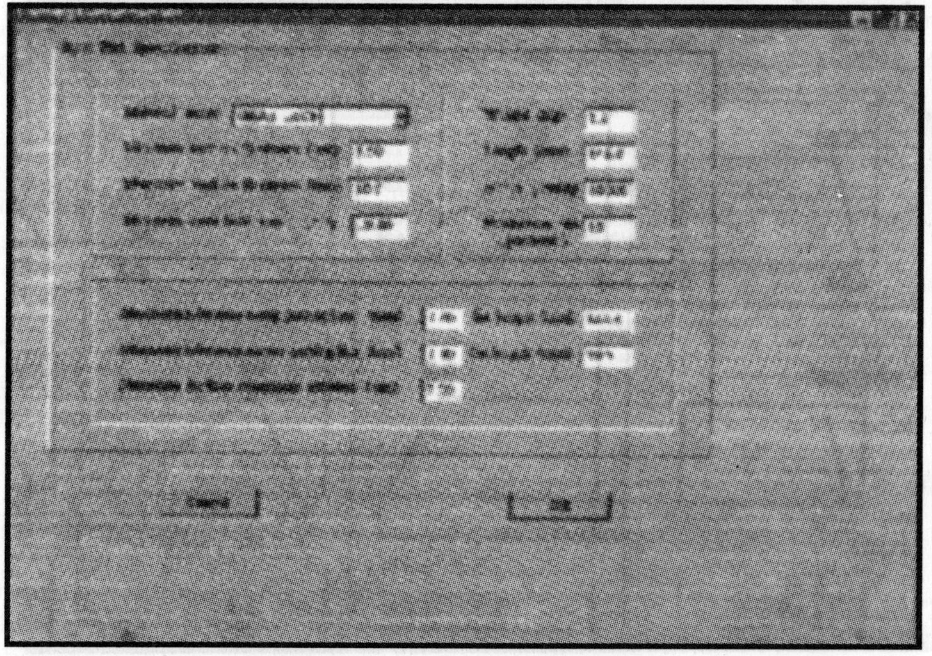

Fig. 27.6. Cast product requirements

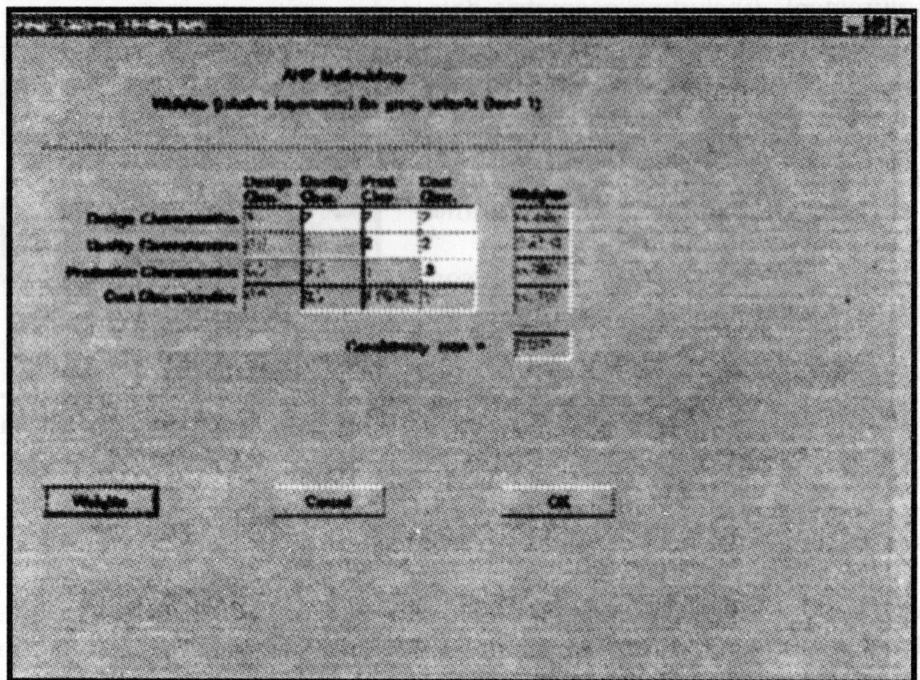

Fig. 27.7. Estimation of weights for group criteria

AHP-FUZZY ANALYSIS FOR CASTING PROCESS RANKING

Group Weight	Group Criteria	Criteria Weight	Detailed Criteria	SAND MOLD	SHELL MOLD	INV CAST
0.39	DESIGN					
		0.25	CORE_HOLE_SIZE	1.00	1.00	1.00
		0.50	SECTION THICKNESS	0.18	0.11	1.00
		0.25	SHAPE_COMPLEXITY	1.00	1.00	1.00
0.71	QUALITY					
		0.43	TOLERANCE_ALONG_PL	0.96	0.00	0.39
		0.35	TOLERANCE_ACROSS_PL	0.94	0.00	0.00
		0.17	SURFACE_ROUGHNESS	0.00	0.51	0.00
		0.05	SURFACE_DETAIL	0.60	0.75	1.00
0.12	PRODUCTION					
		0.31	LEAD_TIME	0.67	0.33	0.33
		0.08	MATERIAL_UTILIZE	0.50	0.63	0.88
		0.19	PRODUCTION_RATE	0.49	0.18	0.83
		0.42	POROSITY VOIDS	0.25	0.25	0.13
0.77	COST					
		0.49	TOOLING COST	0.75	0.38	0.25
		0.28	DIR_LABOUR_COST	0.75	0.50	0.13
		0.07	EQUIPMENT COST	0.75	0.50	0.38
		0.16	FINISHING_COST	0.25	0.75	0.63
			DESIGN	0.55	0.55	1.00
			QUALITY	0.70	0.12	0.22
			PRODUCTION	0.44	0.28	0.23
			COST	0.53	0.48	0.78
			TOTAL SCORE	0.60	0.29	0.34

Fig. 27.8. Compatibility evaluation and ranking

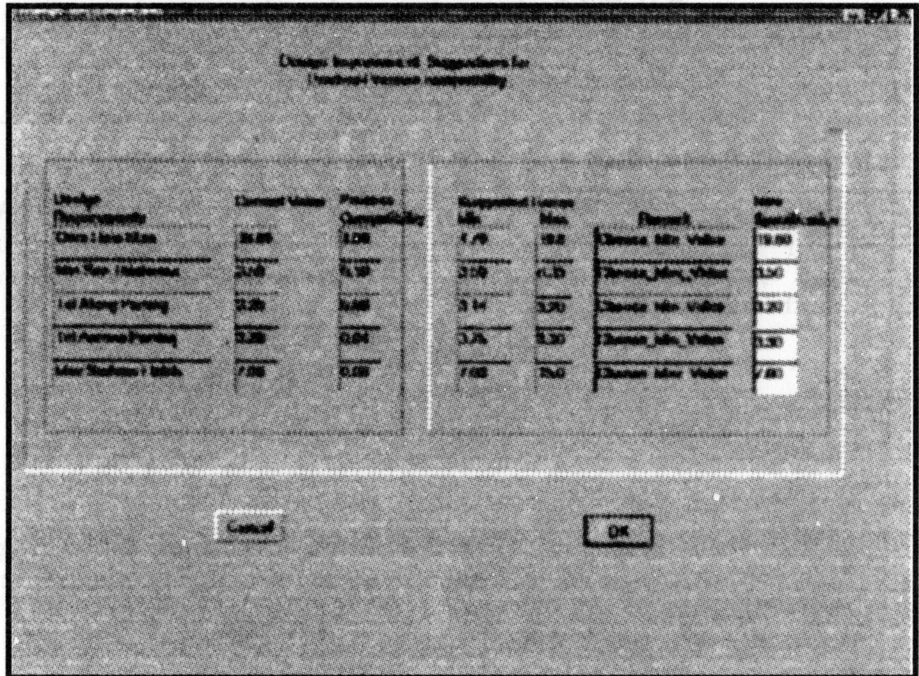

Fig. 27.9. Suggestions for design improvement

A detailed analysis of the product-process compatibility shows that the design group criteria, especially the section thickness criterion, have a low matching with the characteristics of sand moulding process. The programme therefore suggests improvement of such design characteristics to further improve the compatibility (Fig. 27.9). For the test case, the requirement for minimum section thickness (3.5 mm) and surface roughness (7 μm) are just within the capability range of sand moulding process, as indicated by their low compatibility values of 0.1 and 0.08 respectively. The programme suggests increasing their values up to a maximum of 6.35 mm and 15 μm respectively. Even a small increase in their values will significantly improve the product-process compatibility.

In this work, a systematic approach for casting process selection and design improvement has been proposed. This is based on evaluating and maximizing the compatibility between product requirements and process capabilities. Fuzzy logic has been used to capture the imprecise nature of process characteristics, and Analytical Hierarchy Process has been employed to objectively assign weights to evaluation criteria. The approach has been implemented in a computer programme and successfully tested on industrial parts.

Further work involves extending the database to include more casting processes and materials. New criteria can be added. In addition, a separate project to determine the most suitable tooling route and to quickly estimate lead-time and costs is underway, which will enable automatic and objective evaluation of all criteria for casting process selection. With these improvements, the programme will become a useful decision-aid to product engineers for creating robust designs and 'building in' quality at the design stage itself.

Table 27.3. Random consistency index table

Matrix order (n)	1	2	3	4	5	6	7	8	9	10
Random Index (RI)	0	0	0.58	0.9	1.12	1.24	1.32	1.41	1.45	1.49

APPENDIX: ANALYTIC HIERARCHY PROCESS (AHP)

1. Construct a pairwise comparison matrix using a scale of relative importance. Assuming n criteria, the pairwise comparison of criterion i with criterion j yields a square matrix $A\,1_{nxn}$ where, a_{ij} denotes the comparative importance of criterion i with respect to criterion j. In the matrix, $a_{ij} = 1$, when i = j and $a_{ij} = 1/a_{ij}$.

$$A\,1_{nxn} = \begin{bmatrix} a_{11}\ a_{12}\ \cdots\cdots\ a_{1n} \\ a_{21}\ a_{22}\ \cdots\cdots\ a_{2n} \\ \cdots\cdots\cdots\cdots\cdots \\ a_{n1}\ a_{n2}\ \cdots\cdots\ a_{nn} \end{bmatrix}$$

2. Find the relative normalized weight (W_j) of each criterion by (a) calculating the geometric mean of ith row (GM_i) and (b) normalizing the geometric means of rows in the comparative matrix.

$$GM_i = \left[\prod_{j=1}^{n} a_{ij}\right]^{1/n} \quad \text{and} \quad W_j = GM_i / \sum_{i=1}^{n} GM_i$$

3. Calculate matrix A3 and A4, such that A3 = Al* A2 and A4 = A3/A2, where, A2 = $[W_1, W_2 \ldots\ldots W_n]^T$
4. Find out the maximum Eigen value ($\lambda_{min.}$), which is the average of matrix A4.
5. Calculate the consistency index (CI) = ($\lambda_{max.}$ – n)/(n – 1)
6. Obtain the random index (RI), for the number of criteria used in decision making.
7. Finally, calculate the consistency ratio (CR) = CI/RI. Usually, a CR of 0.10 or less is considered acceptable.

28

Automated Rough Turning

N. Popplewell, S. Balakrishnan* and P. Woo**

A major concern in rough turning is to maximize the material removal rate (MRR) without significantly sacrificing the cutting tool life. This paper describes a novel method of utilizing an inexpensive laser sensor for mapping the profile of a rotating workpiece on a lathe regardless of its surface roughness, colour or orientation. An efficient rough turning strategy is then developed for reducing machining time. The material removal rate is observed to reduce it by 36 per cent in comparison to conventional rough cutting. The methodology is also applied to automatically control chatter by adaptively modifying the depth of cut.

INTRODUCTION

The maximization of material removal rate (MRR) is normally achieved by empirically selecting the feed rate and depth of cut based upon the size and material properties of the workpiece as well as the type of cutting tool employed. However, once a selection is made these two parameters are maintained at a constant level throughout the conventional turning. One strategy is to reduce the cutting time by 50 per cent which requires the maximum cutting force to be maintained by adaptively changing the feed rate. A high cutting force reduces the cutting tool's life. A larger depth of cut increases the MRR but chatter is more likely to occur. The depth of cut is usually selected close to the upper limit determined from the lathe's maximum power. In rough turning, the metal removal process can begin with an unknown shape, which must be measured before appropriate cutting parameters can be determined. A laser is often incorporated in devices for mapping the profile of the workpiece.

Experimental Setup

A schematic diagram of the experimental setup is shown in Fig. 28.1. The machine tool is a retrofitted lathe powered by two electrocraft servomotors. The laser device is attached to the lathe's tool post. The major components include a Single phase, 155SL He-Ne laser, a sensor, gear train, rigid arm, two mirrors and a stepper motor. A reorientation mirror redirects the 1 mm diameter beam of the laser before hitting the stepper motor driven oscillating mirror, so as to be swept across a cross-section of the work-piece. The laser beam's movement is synchronized to that of the rigid arm by ensuring that the arm and oscillating mirror have the same center of rotation and a common 2:1 gear train. Consequently, the light beam, when interrupted, is always captured by a sensor held near the arm's free end. The laser beam diverges at two

*Department of Mechanical and Industrial Engineering, University of Manitoba, Winnipeg, Manitoba, Canada.

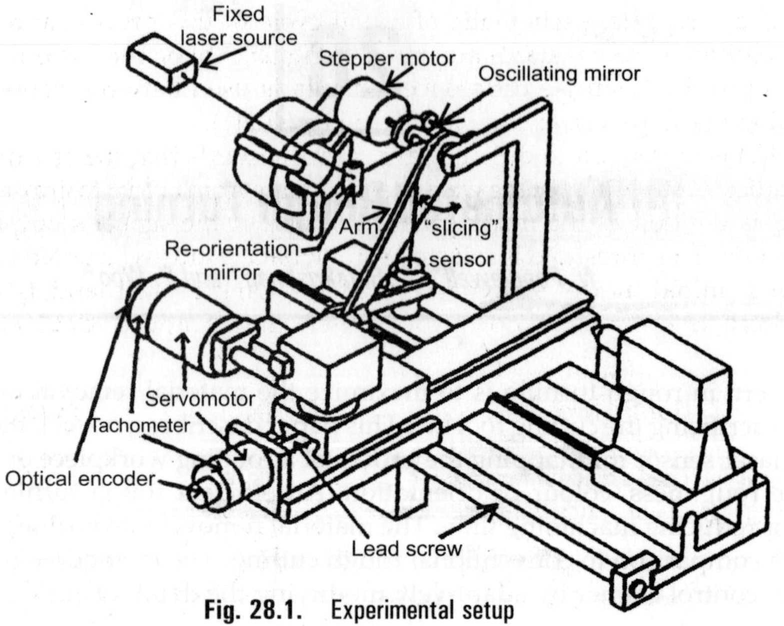

Fig. 28.1. Experimental setup

concave lenses to a cover and a photocell inside the sensor. The photocell behaves as a light controlled resistor that is connected to a voltage divider circuit to produce a voltage that is proportional to the intensity of light received. This intensity, of course, alters and depends instantaneously upon the percentage of the laser beam's cross-section that is blocked by the workpiece. The controller is an IBM compatible computer with three interface cards: one to control the servomotors, the second one to control the stepper motor, and the last one is a 12 bit analog to digital converter.

Profile Mapping

The basic principle of using the proposed laser device is described in this section. Fig. 28.2 shows a rotating bar. Two regions can be seen in the image displayed on horizontal screen below the bar. A gray region, an otherwise bright, is due to the intermittent interruption of the light caused by the rotation of the bar's corner. An interior dark region, conversely, corresponds to a perpetual light blockage by the bar's core. Borders between the dark and gray regions are called "inside" profiles whereas those between the gray and bright regions are termed "outside" profiles. Half the distance between the two outside profiles indicates the cutting tool's minimum starting offset and half that between the inside profiles gives the maximum radius that can be turned into a cylinder. These profiles are detected, in practice, by a change in the light intensity falling on the photocell sensor as it is swept back and forth over a 30° arc beneath the

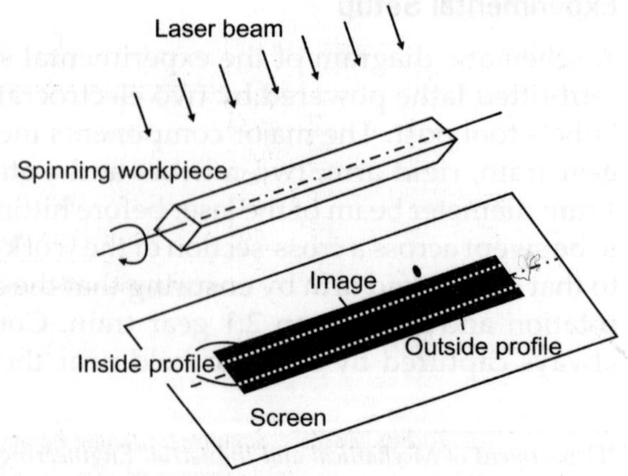

Fig. 28.2. Formation of profiles

bar. Figure 28.3(a) presents a schematic of a half cycle of this process at a particular cross-section of the workpiece. Key instants are noted in Fig. 28.3(b) on the corresponding history of the sensor's output S_d. A simple difference formula is used between consecutive values to derive the analogous slope variations plotted in Fig. 28.3(c).

Figure 28.3(c) is employed to define t_o and t_f, the instants that the sensor's light intensity changes that indicate alterations in the workpiece's image from bright to gray and subsequently to dark. The t_o is defined as the instant that the slope of the sensor's output first crosses a 20 mm/s threshold. This threshold is selected somewhat arbitrarily to avoid false triggering by the 13 mm/s or so initial "noise" shown in Fig. 28.3(c). On the other hand, t_f is the later instant after which there is essentially no output. Then this temporal data is converted into

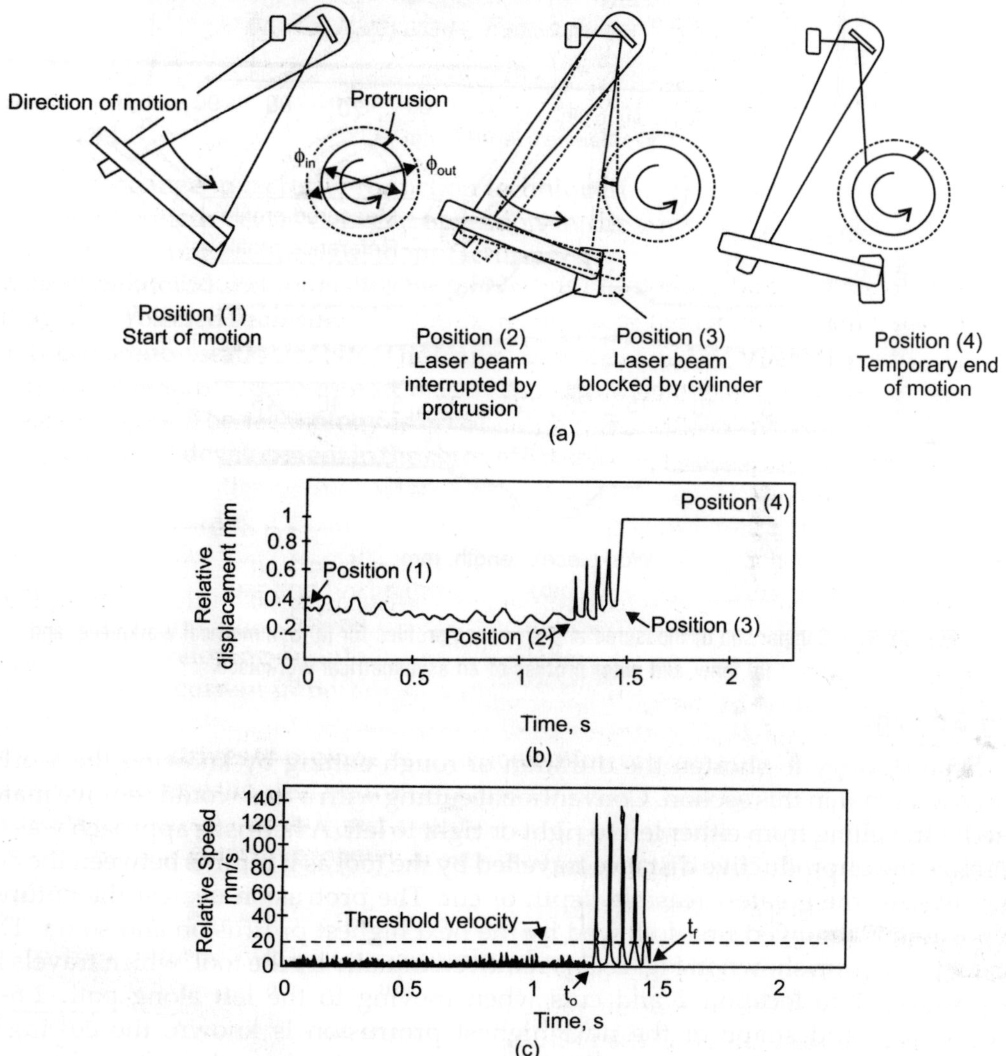

Fig. 28.3. Scan of a cylinder having a protrusion and rotating at 1100 RPM. (a) Locations of arm, (b) Relative desplacement, and (c) Relative speed

corresponding dimensional information by using a calibration curve obtained from seventeen precision made cylinders having different diameters. Figures 28.4 (a) and (b) show the results of scanning a symmetrical and an asymmetrical workpiece. Only the upper half of the symmetrical workpiece, which was machined to an accuracy of +/– 0.03 mm, is shown.

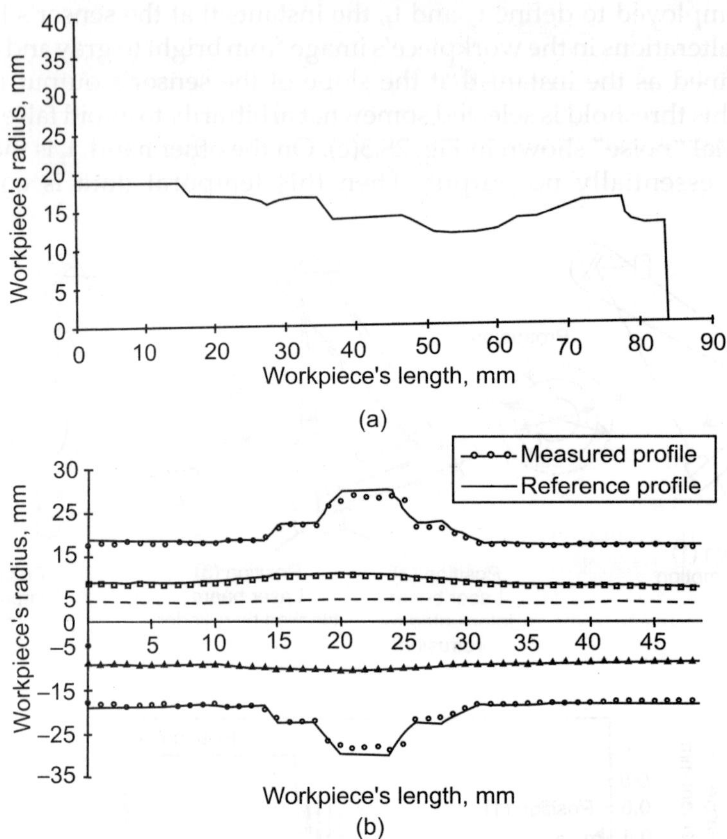

Fig. 28.4. Comparison of measured and reference profiles for (a) Symmetrical workpiece, and (b) Inner and outer profiles of an asymmetrical workpiece

Cutting Strategy

A beneficial strategy to shorten the duration of rough cutting by knowing the workpiece's profile is described in this section. Conventional cutting with a tool would remove material by repeatedly travelling from either left to right or right to left. A heuristic approach was devised to minimize the unproductive distance travelled by the tool as it returns between the cuts. The strategy utilizes the greatest feasible depth of cut. The protrusion nearest the cutting tool's home position is removed first followed by the next highest protrusion and so on. Thus, the rightmost protrusion shown in Fig. 28.5 is removed initially by the tool, which travels from its home position, 1 to location 2 and cuts when moving to the left along path 2-6-3-7-4-8. As the location and shape of the next highest protrusion is known, the cutting can be arranged to relocate to location 9 and to cut progressively along the leftward legs of path 9-13-10-14-11-15-12-16 until the protrusion is also removed. Then, the tool is moved to the predetermined location 5 before performing the last cut and returning home. It was

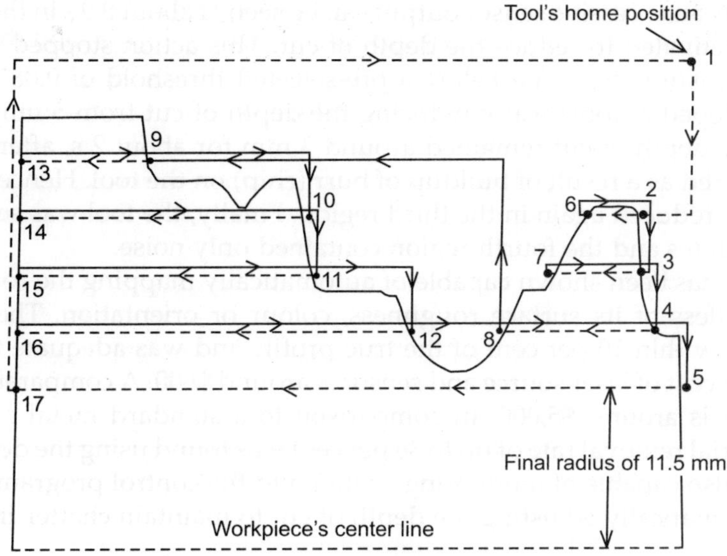

Fig. 28.5. Tool path for profiles shown in Figure 28.4(a)

confirmed that these paths coincided with the shortest non-cutting route generated by a more computationally intensive, branch-and-bound strategy.

Chatter

Chatter is produced when self-excited and excessive vibrations often occur when extreme conditions are encountered in metal cutting. The difference between the sensor's minimum and maximum output $S_{max.}$ was recorded for the strategy to quantify chatter. Four slender carbon steel bars, each having a 25.0 mm diameter and 38.0 mm length, were used to investigate the chatter. Several turning operations with varying depths from 3 to 5 mm were undertaken. A control programme was implemented to automatically reduce the depth of cut when the $S_{max.}$ exceeded a threshold. A representative data is presented in Fig. 28.6. For the four regions

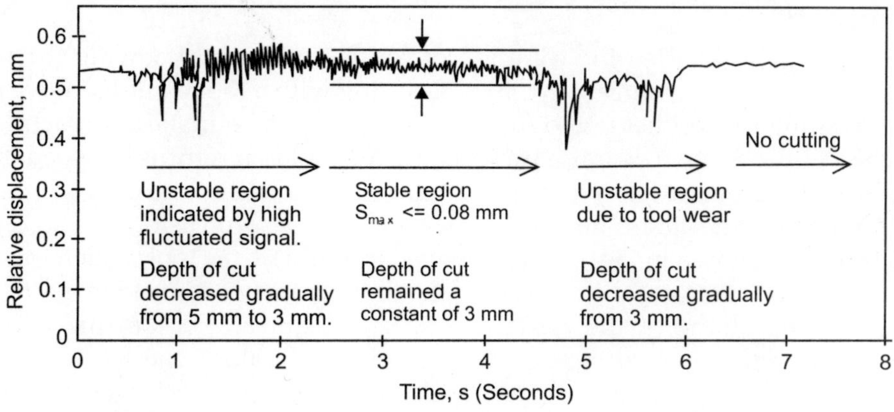

Fig. 28.6. Changes in sensor response when modifying depth of cut

marked in Fig. 28.6, an excessive sensor output can be seen at about 0.9 s in the first region. The servomotor was activated to reduce the depth of cut. This action stopped at approximately 2.5 s when the measured $S_{max.}$ fell below a pre-selected threshold of 0.08 mm. The control programme developed automatically reducing the depth of cut from 5 mm to 3 mm. In the second region, the depth of cut remained around 3 mm for about 2 s, after which excessive vibrations reoccurred as a result of buildup of burr (chip) on the tool. Hence, the depth of cut was automatically reduced again in the third region. Finally, the tool was separated from the workpiece at about 6 s and the fourth region contained only noise.

A laser device has been shown capable of automatically mapping the rotating profile of a workpiece, regardless of its surface roughness, colour or orientation. The accuracy of the profiled map was within 10 per cent of the true profile and was adequate to implement the methodology. The cost of laser source and sensor is around $600. A comparable contact sensor utilized in CMMs is around $5,000. In comparison to a standard metal cutting strategy, a reduction in material removal rate of up to 36 per cent was found using the developed strategy. The technique is also capable of monitoring chatter and the control programme was found to be capable of automatically adjusting the depth of cut to maintain chatter free cutting.

29

Tool Wear in the Machining of Advanced Ceramics

S.C. Laroiya, M. Adithan** and Dell K. Allen****

Advanced ceramics are extensively used in modern technological world. A major breakthrough is required in lowering the production costs and improving producibility and reliability in use for advanced ceramic components and products to achieve their predicted potential. Machining is needed to produce the required size, shape and surface finish not possible by forming methods. To accelerate the commercialization of advanced ceramics low cost machining is of interest to the researchers all over the world.

The study of the tool wear problems is associated with machining methods like diamond machining. Electro discharge machining, ultrasonic machining and rotary ultrasonic machining in the machining of advanced ceramics is important. Experiments have been conducted to machine high alumina, silicon carbide and steatite ceramic, using ultrasonic drilling process and the data thus generated has been analysed and inferences drawn. Production accuracy, surface roughness, machining rates and tool wear values obtained while machining advanced ceramics are discussed.

INTRODUCTION

Ceramics have been in use since time immemorial. Advanced ceramic applications presently include mechanical components, hip joints, human body parts, computer chips and parts of spaceships. Their application is of special interest to the researchers all over the world because of their excellent properties like high strength-to-weight ratio and resistance to heat, wear and corrosion. During the sintering of these materials shrinkage cannot be entirely ruled out and therefore machining is needed to attain proper accuracy, shape, etc. Since the development of ceramic engines most of the ceramic components have been machined with diamond machining methods. The main aim of the scientists and researchers these days has been the development of low cost machining of advanced ceramics.

Important properties of various ceramics are that they are weak in tension but strong in compression, e.g.. alumina ceramics has tensile strength of 138 MPa and compressive strength of 240 MPa; homogeneous ceramics of single phase are usually stronger than those with several phases; ceramics possess good creep resistance, high corrosion resistance, high thermal shock resistance as well as the ability to withstand high service temperatures of the order of 1650°C. Amongst the new generation ceramics the most widely used ceramics in the industry

* *Technical Teachers' Training Institute, Chandigarh, India.*

** *Vellore Institute of Technology, Deemed University, Vellore, India.*

*** *Brigham Young University, Provo, Utah, U.S.A.*

are SiC, Si_3N_4, CBN, Al_2O_3, UCon. (a proprietary Nitrided refractory metal alloy developed by Union Carbide, USA).

Differences in the type of atomic bonding are responsible for hardness and Young's modulus of ceramic materials. Materials with metallic bonding exhibit high ductility, covalently bonded ceramics typically are hard, strong and have high melting temperature.

NEED FOR MACHINING OF ADVANCED CERAMICS

There are many reasons why we should go in for machining of advanced ceramics after their compacting and sintering:

(a) During the sintering process, shrinkage of the material cannot be entirely avoided; hence to achieve closer tolerances and dimensional accuracies of parts machining of ceramics is required;
(b) Slots, grooves and side holes are required to be produced as a secondary operation after compacting and sintering since these configurations are difficult to obtain during compacting and moulding;
(c) Manufacturing process has to accommodate limitless diversity through holes, blind holes, slots, shapes and odd contours for production of complex and intricate parts;
(d) For slicing into small pieces and their wafers.

METHODS AVAILABLE FOR ASSOCIATED MACHINING OF ADVANCED CERAMICS AND TOOL WEAR PROBLEMS

Diamond Turning

While it is possible to machine the advanced ceramics with sintered diamond tools, the tool wear is greatly affected by the cutting pressure and consequently limits the feed rate. Tool shape should be so designed that the cutting pressure or the uncut chip thickness is uniform along the cutting edge, i.e. a pointed tool with a larger side cutting edge is desirable.

Diamond Grinding

For grinding of ceramics grinding machines with a high stiffness of 10 N/µm. between the tool holder and the workpiece spindle are required. Machines designed to grind various types of steel with CBN wheels may also work in certain ceramic machining applications. For ceramic components machining higher wheel speeds up to 100 m/sec. are required to improve material removal rates and surface finish.

The mechanism of material removal in diamond grinding of ceramics is primarily due to the micro brittle fracture of the material. The diamond grinding process is completed in two phases, viz. roughing and finishing. A high workpiece speed and a small depth of cut are desirable for roughing process to reduce the grinding force, while the workpiece speed should be kept relatively low in the finishing process to overcome the chances of getting a damaged surface layer.

The problem of utmost importance in diamond grinding of ceramics is the excessive increase in the grinding force with time elapsed and the occurrence of chipping and cracking in the workpiece.

A new dressing technology has been developed for machining of SIALON (Si-Al-O-N) ceramics which is difficult to cut by conventional diamond machining. This is because the diamond wheel looses its cutting ability with time, due to pile up of ceramic chips around the diamond grains. Newly developed dressing technology consists of continuous removal of the chips during the cutting off process, in which a glass fiber reinforced plastic plate is used as dresser. Sialon ceramics can be cut off under stable conditions at higher cutting speeds with a smaller cutting force. This also results in less wear of the diamond wheel.

Creep Feed Grinding

Creep feed grinding has an important role in minimizing the chipping of the workpiece which quite often occurs at the entry and exit of the grinding wheel. It is therefore essential to introduce continuous dressing of grinding wheel to prevent the increase in the grinding force with time to take advantage of creep feed grinding.

Resin bonded wheels wear away faster but these produce much lower grinding forces while machining high Alumina ceramics. Selection of an appropriate grinding wheel and cutting fluid are essential for efficient grinding of advanced ceramics.

From the point of view of surface roughness and wheel wear emulsifiable oil shows better performance than other chemical fluids.

Toyota Machine Works at Japan conducted several experiments to improve the quality of surface produced while grinding of ceramics. Nickel coated friable abrasives used in resinoid bonded grinding wheels produced better results while grinding high Alumina ceramics. However, while grinding Silicon Nitride ($Si_3 N_4$) components it was found that the material strength was reduced after grinding.

In creep feed grinding a combination of a stiff machine tool, which eliminates vibrations, with high positioning accuracy and high rotational speed results in improved efficiency and dimensional accuracies.

In Si/SiC components, surface grinding should be carried out based on an analysis of process kinematics and a study of the basic mechanisms of material removal because such methods will lead to considerable increase in material removal rate, component strength and tool life.

Milling

For milling of machinable ceramics, Tungsten Carbide WC tools are required and the chip size is determined by the relationship between the angle of work corner and disengagement angle in cutting. Cutting tool with large nose radius provides high efficiency cutting and increases tool life while decreasing the cross-sectional area of uncut chip increases specific cutting pressure and thereby cutting forces.

Ultrasonic vibrations to a grinding wheel reduces grinding force and also a combination of vibrations and EDM is possible for machining certain types of structural ceramics.

ELECTRO-DISCHARGE MACHINING OF CERAMICS

EDM process, when used on complex Zirconium Diboride ceramic components, had higher material removal rates as compared to diamond grinding which can result in cost savings. Fewer setups and operations were required with the EDM process.

It is possible to machine ceramic materials by EDM process if electrical conductivity is of the order of 100 Ohm-cm. EDM achieves high metal removal rate as compared to traditional techniques like diamond machining methods. However, the metal removal mechanism for machining conductive ceramics by EDM differs considerably from those involved in normal metal machining.

LASER BEAM MACHINING OF CERAMICS

Laser beams can be used as a tool for machining of advanced ceramics. The technique has high productivity and eliminates the problems of tool wear, but process optimization is required for its extensive application. There is a reduction or total absence of cracks and other defects in CO_2, Nd, YAG and excimer laser drilling of ceramics with short pulses and high beam intensities.

HOT MACHINING OF CERAMICS

The cutting mechanism of the ceramics changes from brittle fracture to plastic deformation with the increase in temperatures, and at high temperatures the machining of ceramics is similar to conventional metal cutting. In hot machining, the surface finish and tool life are improved remarkably. In hot machining, ceramics can be machined with high metal removal rates and with good surface integrity.

WATER JET MACHINING OF CERAMICS

For ceramics machining, a high velocity fluid jet either with or without abrasive particles in the cavitating or in continuous mode can be used. This type of machining is best suited for cutting slots and grooves or for trepanning large holes in ceramics. There are no heat effects nor large mechanical forces in the cutting process. Consequently, very fragile ceramics can be cut without any damage. A high pressure abrasive water jet can be used for machining certain varieties of high alumina ceramics. The system consists of a robotically controlled water jet that handles pressures up to 200 MPa with water feed rates (flow rates) ranging from 0.10 to 1.30 kg/minute.

CHEMICAL ETCHING OF CERAMICS

Ceramics, such as silicon wafers and quartz crystals, can be machined with chemical etching and this results in the production of good quality holes. However, strict and extensive controls were needed in the use of toxic chemicals and the process is very time consuming.

ULTRASONIC MACHINING OF ADVANCED CERAMICS

Since ultrasonic machining is a non-chemical, non-electrical and non-thermal process, it provides the user with a capability of machining complex shapes and contours in many hard and brittle materials.

Ultrasonic drilling process is preferred over laser drilling. In laser drilling, there is always the drawback of microcracking and presence of laser slag. Hence, ultrasonic machining is always preferable for high quality, high yield and high reliability applications.

For materials like structural ceramics and glasses, the structure of the material is such that large forces are required to machine these materials and these cause localized surface cracking which leads to reduced strength and load carrying capacity of the components. The strength properties of the components can be retained by careful machining using USM. Ultrasonic machining provides means by which ceramics can be machined successfully preserving their surface integrity.

A recent application of ultrasonic machining is the machining of three-dimensional shapes in new generation ceramics. This process is especially suitable for machining of complex three-dimensional shapes up to 100 mm in diameter in brittle materials. USM has a number of important advantages for ceramic applications: conducting and non-conducting materials can be machined irrespective of their hardness; there are no chemical or electrical alterations on the workpiece surface and complex three-dimensional contours can be machined as quickly as simple ones. However, the main problems associated with successful application of ultrasonic machining are:

- Low material removal rates and hence low productivity;
- Likely production inaccuracies of the holes machined;
- Presence of longitudinal and lateral tool wear; and
- Gradual wear of the abrasive particles themselves.

Material removal rate in USM is influenced by a number of parameters such as amplitude and frequency of vibrations, static load, type and size of the abrasives, concentration of the abrasives in the slurry, method of feeding of abrasive slurry, the medium of the slurry, the work material properties, tool geometry and tool material.

ROTARY ULTRASONIC MACHINING (RUSM) OF CERAMICS

Various ceramic materials, such as alumina, ferrite and porcelain, can be successfully machined preserving their surface integrity. Rotary USM is superior to conventional USM in all respects. The rotary mode yields higher material removal rates, lower surface roughness and better accuracy. Rotary USM is always recommended for machining axisymmetric holes.

INVESTIGATIONS CARRIED OUT

The advanced ceramics used in the study are Alpha SiC, High Alumina (96 per cent) and steatite ceramics. Experiments have been conducted on a vertical 600 Watts ultrasonic drilling machine manufactured by Kerry's Ultrasonics, U.K. The abrasive: water ratio in the slurry used is 1:4 (by weight); SiC 280 Mesh abrasives are used in the investigations. For tracing and computing tool wear and production accuracy a profile projector is used.

Production Accuracy in the Machining of Advanced Ceramics

Due to external feeding of the slurry and as the point of feeding is constant, gradual rounding off of hole at entry occurs and the hole size at entry tends to be enlarged. As drilling continues and proceeds further, there is a tapering effect on the hole due to the onset of lateral wear of the tool.

Conicity or taper in the hole is mainly due to lateral wear of the tool; use of tougher and harder tool materials, such as stainless steel and tungsten carbide, reduces lateral wear of the tool and the conicity of the holes produced. Conicity also depends on the nature of work material machined.

Out-of-roundness (OOR) is caused by the non-uniform flow of abrasive particles in the gap between the tool and the workpiece; these particles scour the tool surface and produce lateral or side wear. The extent of lateral wear depends on the rate of abrasive flow and as this is not uniform around the circumference of the tool, tool wear is also not uniform around the circumference of the tool. Out-of-roundness of the tool occurs first, followed by out-of-roundness of the hole produced. Out-of-roundness of the holes produced is assessed by diameter measurement. It is observed that out-of-roundness of the hole is more at entry. Production accuracy of holes machined is shown in Tables 29.1 and 29.2.

It has been observed that as the depth of the hole increases the machining rate decreases for a while and then remains almost a constant rate (Figs. 29.1 and 29.2). This is due to the non-availability of the fresh abrasive slurry at higher depths of hole.

PRODUCTION ACCURACY OF HOLES MACHINED

Table 29.1. Work Material: High Alumina

Sr. No.	*Diameter of hole (mm)*	*Oversize of hole (mm)*	*Out-of-roundness at entry (mm)*	*Out-of-roundness at exit (mm)*	*Conicity **
1	2.90	0.15	0.18	0.32	0.05
2	3.00	0.18	0.45	0.53	0.06
3	3.50	0.22	0.19	0.20	0.09
4	4.50	0.16	0.23	0.28	0.07
5	5.00	0.10	0.36	0.41	0.08

* Conicity or Taper for 4 mm length of hole

Table 29.2. Work Material: Alpha Silicon Carbide

Sr. No.	*Diameter of hole (mm)*	*Oversize of hole (mm)*	*Out-of-roundness at entry (mm)*	*Out-of-roundness at exit (mm)*	*Conicity **
1	2.500	0.255	0.150	0.230	0.385
2	2.900	0.110	0.120	0.130	0.005
3	3.500	0.140	0.210	0.210	0.020
4	4.500	0.280	0.270	0.280	0.015
5	6.500	0.185	0.125	0.175	0.075

* Conicity or Taper for 5 mm length of hole

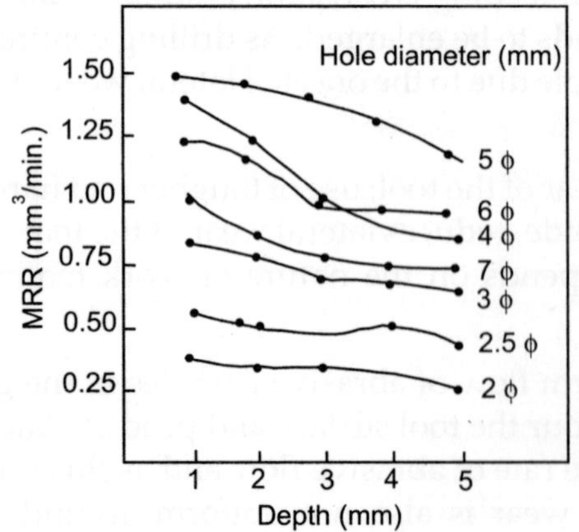

Fig. 29.1. Material removal rate vs depth. Work material: Alpha silicon carbide, Tool material: Stainless steel

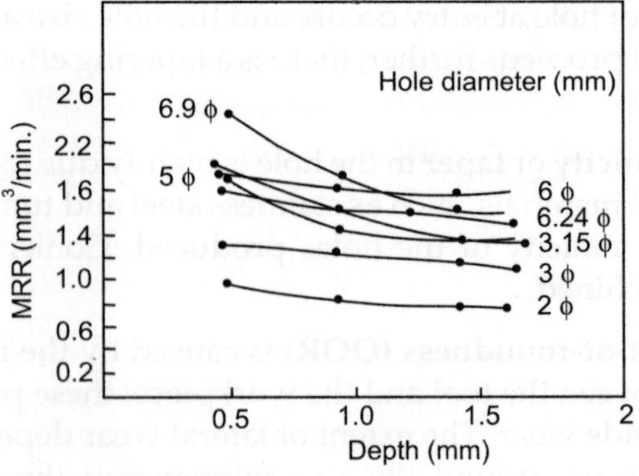

Fig. 29.2. Material removal rate vs depth. Work material: Steatite, Tool material: Stainless steel

Tool Wear in the Machining of Advanced Ceramics

Tool wear is an important factor which affects the performance of the ultrasonic drilling process. Tool wear results in frequent tool replacement. As the tool length and weight decrease the tool is not in resonance with the driving oscillator, amplitude of vibrations reduces and the machining rate decreases. A contoured tool profile results in dimensional inaccuracy of the holes machined. Tool wear also leads to taper formation in the holes resulting in poor circulation of the abrasive slurry and reduced machining rate. Tool wear in ultrasonic drilling is influenced by a number of factors, such as cutting time, static load, type of abrasives used and tool material which, in turn, all affect the performance of the process considerably.

Longitudinal tool wear is significant only at depths greater than 15 mm while machining ceramics. Relative tool wear is about 45 times more while machining tungsten carbide when compared to machining of glass. Tungsten carbide is difficult to machine because of its higher toughness when compared with that of glass. All the parameters which affect the material removal rate also equally affect the tool wear rate. In addition to longitudinal wear and lateral wear, a third type of wear known as cavitation wear also exists. This is confirmed by SEM analysis of worn out tool surface (Fig. 29.3). Tool wear in the case of machining of Alpha Silicon Carbide was found to be several times more as compared to machining of high Alumina (96 per cent purity).

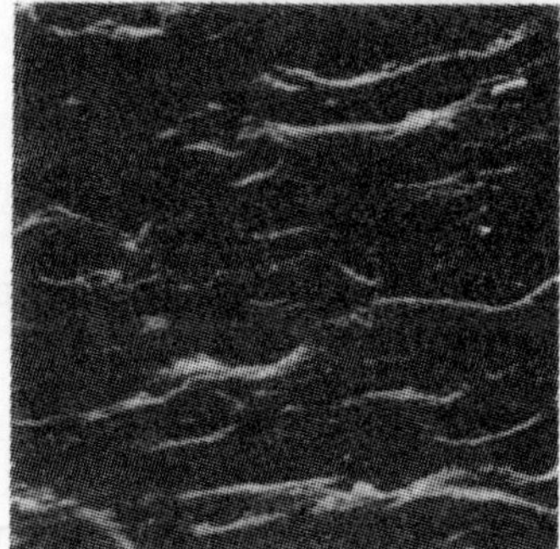

Fig. 29.3. Scanning electron micrograph of tool surface produced after drilling holes in high alumina. The surface shows cavities formed due to cavitational wear (4000 x) (Tool material: Stainless steel)

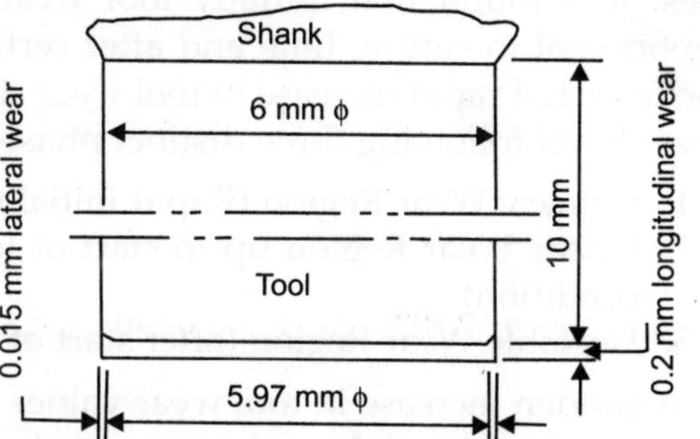

Fig. 29.4. Schematic representation of tool wear (Work material: Steatite) (Tool material: Stainless steel)

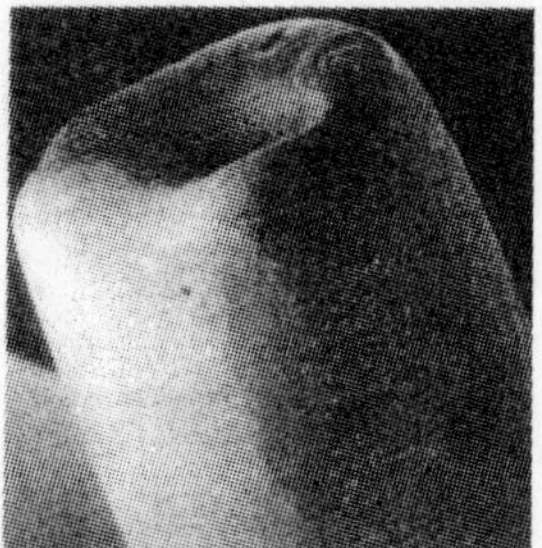

Fig. 29.5. Scanning electron micrograph of wear profile of tool (40 x) (Work material: High alumina) (Tool material: Stainless steel)

Tool Wear Propagation and Production Accuracy

For studying the pattern of tool wear propagation the shank was unscrewed from the transducer after the drilling operation and tool end viewed under a profile projector with a high magnification (× 100). The change in the profile of the tool after drilling every hole was traced. Both longitudinal and lateral wear values were measured from a preset reference surface to an accuracy of 1 micron. The changes of the tool contour from hole to hole become evident through the recording of the successive longitudinal sections into one another. From this the development of the wear profile could be seen Fig. 29.6 shows a plot between cumulative longitudinal tool wear and cumulative cutting time when Alpha Silicon Carbide is machined with different diameter holes. It is found that initially tool wear is proportional to cutting time and after certain time there is a rapid increase in tool wear. The curve shows following three distinct phases:

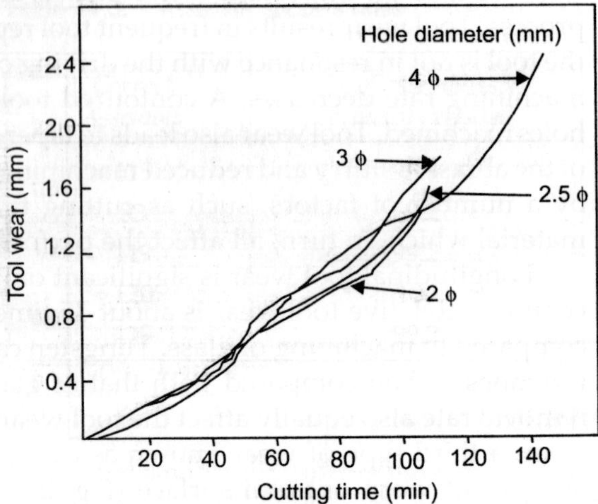

Fig. 29.6. **Tool wear vs cutting time (Work material: Alpha silicon carbide) (Tool material: Stainless steel)**

1. Primary Wear Region (Rapid initial wear)
2. Linear Wear Region up to start of tool breakdown (lower rate of wear during stable condition)
3. Unstable Wear Region (after start of tool breakdown until tool failure).

A sudden increase in tool wear values at the end of linear wear region confirms that the start of the tool breakdown has occurred after a certain definite time and further machining results in a steep increase in tool wear leading to catastrophic tool failure.

The value of surface finish of the holes drilled in SiC varies from 0.5 to 0.7 μm (Ra).

30

Cost Analysis in Concurrent Engineering Design *vs.* Sequential Engineering Design

Robert C. Creese and L. Ted Moore**

INTRODUCTION

Concurrent engineering has been credited with numerous successes in improving quality, reducing costs, decreasing the product development cycle time, and reducing scrap and rework for several companies in the USA. The reported benefits attributed to concurrent engineering are:

1. Improved design quality which resulted in reduction in engineering change orders of more than 50 per cent.
2. Product development cycle reduction by as much as 40 to 60 per cent less than sequential engineering.
3. Manufacturing cost reduction by as much as 30 to 40 per cent.
4. Scrap and rework reduced by as much as 75 per cent.

There are also numerous specific instances of savings attributed to concurrent engineering as reported by the Institute for Defense Analysis, U.S.A. Some examples are: Aerojet Ordinance, U.S.A. which identified the correct design parameter settings on the process for making pyrotechnic pellets and obtained 100 per cent yield on 54 production runs. This was the first time in the history of the product. A 100 per cent yield was observed in a reasonable time period. AT&T, U.S.A. increased its acceptance at the first production run from 50 to 90 per cent. At Boeing, U.S.A. parts and material lead time has been reduced by 30 per cent and material shortages have been reduced from 12 per cent to less than 1 per cent. Deere and Company, U.S.A., reduced developmental time for new products by 60 per cent which resulted in a cost savings of 30 per cent. Other examples have been reported from industries, such as Grumman, Hewlett-Packard Company, IBM, ITT, McDonnell Douglas, Northrup, and Texas Instruments, U.S.A.

Concurrent engineering is a management and engineering philosophy intending to improve quality, decrease costs, and reduce the lead time from concept through development to production for new products. The philosophy of concurrent engineering is not new and the terms "systems engineering", "design for manufacturing", "flexible manufacturing", "rapid prototyping", "computer integrated manufacturing", "integrated cross-functional engineering",

**Industrial Engineering Department, West Virginia University, Morgantown, West Virginia, USA.*

and others have been used to describe approaches similar to concurrent engineering. Concurrent engineering is more comprehensive than these approaches in that it emphasizes management as well as engineering skills.

The **definition of concurrent engineering** developed by the Institute for Defense Analysis, U.S.A., is:

Concurrent Engineering is a systematic approach to the integrated, concurrent design of products and their related processes, including manufacturing and support. This approach is intended to cause the developers, from the outset, to consider all elements of the product life cycle from conception through disposal, including quality, cost, schedule, and user requirements.

The traditional approach to product development and production is frequently referred to as sequential engineering, emphasizing a predominantly unidirectional information flow, whereas concurrent engineering has multidirectional information flows. Unidirectional flow has evolved as companies have increased in size and delegated responsibilities to different departments. These departments are concerned primarily with their specific task and do not concern themselves with the overall product for various reasons.

Some of this over specialization is a result from management's fear that "spies" from other companies will steal their "secrets", which fear, shall result in attitudes of "it is none of your business, it is their problem". Smaller companies have tended to practice concurrent engineering as a natural mode of communication between departments. The information flow differences between sequential and concurrent engineering are illustrated in a general fashion in Fig. 30.1.

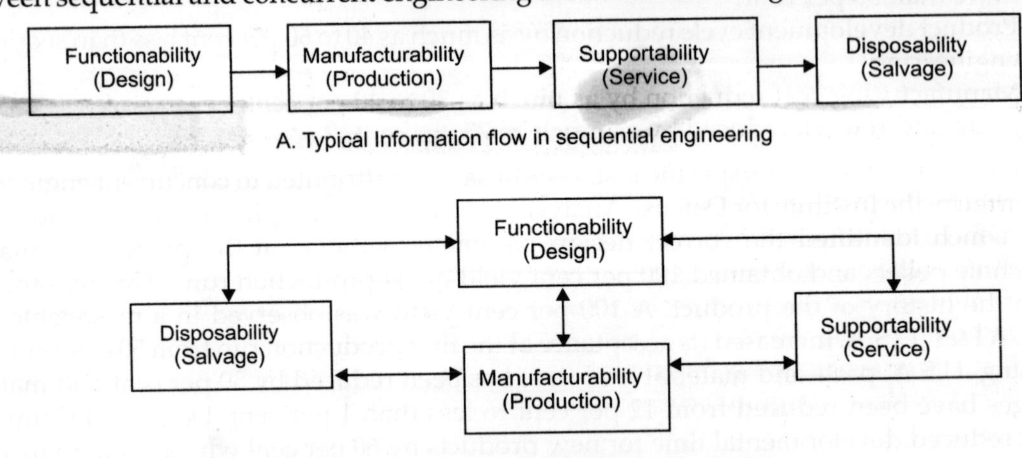

Fig. 30.1. Information flow in sequential and concurrent engineering

CONCURRENT ENGINEERING AS A MANAGEMENT PHILOSOPHY

Concurrent engineering can be compared to the topics of group technology (GT) and total quality management (TQM) in that all three are considered to be a "philosophy". Group technology (GT) has been defined as:

Group Technology (GT) is a manufacturing philosophy which identifies and exploits "the underlying sameness" of parts and the manufacturing processes.

One of the recent definitions of total quality management (TQM) is:

Total quality management (TQM) is a leadership philosophy, organizational structure, and working environment that fosters and nourishes a personal accountability and responsibility for quality and a quest for continuous improvement in products, services and processes.

These definitions indicate the importance of the "philosophy" element and suggest the difficulty of formulating a specific, detailed definition. Concurrent engineering, like group technology (GT) and total quality management (TQM), is not a specific tool, architecture, technique or process. One of the major impediments to the implementation of concurrent engineering is the failure of management to recognize that it is a management and engineering philosophy which must be implemented at the top levels of the organization before it can be successfully implemented throughout the corporation. Top management has generally failed to comprehend and/or execute the leadership needed for the implementation of concurrent engineering. The Institute for Defense Analysis, U.S.A., stated that successful implementation of concurrent engineering required top-down implementation through leadership rather than through dictates.

The differences between concurrent engineering, group technology (GT), and total quality management (TQM) are difficult to delineate. The initial perceived distinction is that total quality management (TQM) is the most comprehensive followed by concurrent engineering and then group technology (GT). This concept fails in that Deere and Company U.S.A., one of the leaders in concurrent engineering, refers to its programme as a "Group Technology System" and uses it to meet 100 per cent quality and to be cost competitive. All of these terms-concurrent engineering, total quality management (TQM), and group technology (GT)-require integration of design, manufacturing, and support for the purposes of reducing cost, reducing cycle time, improving quality, and reducing product development time. Real distinctions between these terms have not been established or accepted at this time.

Since concurrent engineering is a recent topic, there are several misconceptions about it and some of the common misconceptions are:

1. Concurrent engineering is not a magical formula for success. It is an approach for improving the efficiency of good people who work hard, but it provides no guarantees of success. It cannot compensate for a lack of talent, but it can make more efficient use of your existing talent.
2. Concurrent engineering, in isolation, cannot produce the type of improvements required. It is part of an integrated corporate competitiveness plan and requires leadership from the top.
3. Concurrent engineering entails simultaneous design of the product and downstream processes, but does not entail design of the product simultaneous with execution of the process. Completion of all design efforts prior to production initiation is required to prevent downstream production problems and delays.
4. Concurrent engineering is not the elimination of phases in the sequential feed-forward engineering process, but it integrates interaction of design, production, support, and disposal requirements for an overall cost-effective product.

Concurrent engineering is a management and engineering philosophy and it will be implemented differently for different companies because of differences in leadership, products, processes, employee talents, suppliers, and customers. Although this makes it more difficult to implement and forces management to lead rather than copy, the results can be significant as indicated initially. It takes time and is costly to install, but concurrent engineering is essential if one is to be competitive in the international market.

COST MODELLING AND CONCURRENT ENGINEERING

Cost is one of the major items in evaluating the success of applying concurrent engineering to various projects. The importance of cost modelling in the design stage has been reported by many researchers. It has been stated by numerous authorities that the designer commits 75 to 85 per cent of the eventual product costs and he is responsible for about 60 to 70 per cent of the unnecessary costs. The unnecessary costs have been estimated to be approximately 20 to 30 per cent of the total product costs. Thus, significant cost savings can be realized when the designers are able to evaluate their designs on a cost effective basis. The purpose of cost models at the design stage is to help the designer reduce, and hopefully eliminate, unnecessary costs under his control. The development of effective cost models is essential to the long-term success of the concurrent engineering concept.

Design has been primarily concerned with the manufacturability of products and has little concern with supportability (repairability and maintainability) and disposability. However, if there is only one customer, such as a government, and they require that the products have certain supportability and disposability characteristics, the producer must include these characteristics in their design and cost models. Thus, military products, such as airplanes, ships, and tanks, have more requirements in the areas of supportability and disposability than most commercial products. For these reasons, little modeling work has been done or reported for supportability or disposability costs.

Numerous types of cost models have been developed, and it is difficult to compare them since they have been developed for different purposes, different companies with different cost structures, and for different products. Cost models have been developed for single components, such as a turbine blade; for assemblies, such as an aircraft engine; and for products, such as a fighter plane. In the models developed for assemblies and for products, parametric cost estimation has proven to be successful whereas for single components the costs of the specific manufacturing operations are used.

A recent classification system for cost models is based upon the different stages in the design. The three levels of design considered are:

1. Conceptual design
2. Preliminary design/design
3. Detailed design

The **conceptual design stage** is that at which the geometry and materials are unknown, except in the case of parts for which they dictate essential product functions. At this stage, the processes are not specified and concerns of supportability and disposability are not considered unless they are essential design considerations. In the conceptual design stage, the most important factor is the functionability, and once that is assured then the concerns of manufacturability, supportability, and disposability are considered. The cost estimations used are an-order-of-magnitude estimate, that is, a minus 30 to plus 50 per cent accuracy.

The cost estimations can be for either an assembly product or a single item product. The cost estimations generally used are the factor method, the function method, and the material cost method. The factor method used factors from historical, measured, or published data and is used for plants or processes rather than for manufactured products. The function method used mathematical expressions with constants and parameters derived for specific processes (welding, turning, rolling, gearwheels, moulding) or for specific classes of parts (weight, size, alloy, and other restrictions). Parametric cost estimating is generally considered a function method, but some of the detailed parametric models can be used for the preliminary design/design models as well as for the conceptual models. The material cost method is used for complex objects which have several parts or operations, such as passenger cars or diesel engines. The material cost method is frequently used for the estimation of castings, and software programmes have been developed to assist in determining the weight of complex castings. General design rules can be applied at this stage, such as "decreasing the number of component parts is advantageous" or "circular holes in sheet metal parts should be larger in diameter than the sheet metal thickness".

In the **preliminary design/design stage** the types of materials are specified and the product is completely dimensioned. Cost estimates in this stage are preliminary, that is, can be used to estimate budgets for manufacturing, and would be in the minus 15 to plus 30 per cent. Costs can be predicted by comparing the product with classes of existing products, calculated from databases, or by using more detailed cost function methods. During this stage design, changes can be evaluated on both the product and the production processes used to manufacture the product. It is during the conceptual and preliminary design/design stages where the designer can make his greatest impact on cost reduction. Some general design rules would be "select allowances and surface finish values to promote the acceptability of the widest allowances and the acceptable surface finish values" or "maximize the amount of standardization".

In the **final detailed design stage,** all the information about the product is known, such as the results of prototype performance and the product design is complete; the specific process operations and process parameters are known; the requirements for product support services and product reliability are known in detail; and the product disposal requirements and related design effects are known. A detailed cost estimate can be made and the results should be within minus 5 to plus 15 per cent. The industrial engineering production details, such as material and process standards, scheduling, production and inventory controls, quality and process controls, production, inspection and testing, product reliability, overhead rates, etc. would be used in the detailed cost estimate.

If the designer is to be able to evaluate designs on a cost effective basis, then cost models must be developed to help the designer. Since the greatest opportunity for savings is in the early design stages, cost models for conceptual design and preliminary design/design will be of most important to the designer. The "Design to Cost" concept emphasizes the importance of making design converge on cost instead of allowing cost to converge on design. The concurrent engineering and design-to-cost approaches require integration of manufacturing, support, and disposal requirements in the early design stages, especially if they are essential design considerations. Cost models need to be developed to include all essential design considerations in the conceptual models, including disposability, supportability, and manufacturability as well as functionability.

CONCURRENT ENGINEERING RESEARCH CENTER

The need for cost-effective, quality, on-shore manufacturing capabilities is of a major concern to the U.S. Department of Defence and the reported successes of concurrent engineering in U.S. manufacturing was of significant interest. One of the efforts to advance concurrent engineering is the sponsoring of an initiative in concurrent engineering by the U.S. Defence Advanced Research Projects Agency (DARPA). This DARPA initiative in concurrent engineering is referred to as DICE. One major component of this initiative is the establishment of a Concurrent Engineering Research Center (CERC) at West Virginia University, U.S.A. The major objective of DICE is to develop a new design and manufacturing technology to improve the competitiveness of American products in international markets. This will be accomplished by

1. Reducing the elapsed time between concept and production of new products.
2. Improving the product quality by "getting it right the first time."
3. Decreasing cost of manufacture by better utilization of resources and optimal product designs.

There are a wide variety of concurrent engineering research activities being pursued at CERC, and some of these are:

- Formulating new knowledge in processing technologies and advanced materials;
- Hosting national conferences and workshops in concurrent engineering and manufacturing productivity:
- Publishing an International Journal of Concurrent Engineering;
- Developing a model curriculum for a course on "Concurrent Engineering"; and
- Hosting visiting scholars from academia and industry.

In addition to the CERC facility, a Consortium of academic and research laboratories has been established to ensure the research focuses on industrial needs and facilitates rapid technology transfer. The Consortium consists of five universities and six industrial research laboratories to focus on research in concurrent engineering.

Cost modelling is one of the critical elements in concurrent engineering. Cost models must now include concerns for disposal, support, and manufacturing as well as for functional design. These concerns must be included in the conceptual cost model and in the preliminary design/design cost model as well as in the detailed cost model. Only when these concerns are appropriately addressed in the cost models can designers fully evaluate their designs on a cost basis. The avoidance of the 20–30 per cent typical unnecessary costs is required if products are to be competitive in international markets.

Index